T0134703

Sustainable Civil Infrastructures

Sustainable Infrastructure impacts our well-being and day-to-day lives. The infrastructures we are building today will shape our lives tomorrow. The complex and diverse nature of the impacts due to weather extremes on transportation and civil infrastructures can be seen in our roadways, bridges, and buildings. Extreme summer temperatures, droughts, flash floods, and rising numbers of freeze-thaw cycles pose challenges for civil infrastructure and can endanger public safety. We constantly hear how civil infrastructures need constant attention, preservation, and upgrading. Such improvements and developments would obviously benefit from our desired book series that provide sustainable engineering materials and designs. The economic impact is huge and much research has been conducted worldwide. The future holds many opportunities, not only for researchers in a given country, but also for the worldwide field engineers who apply and implement these technologies. We believe that no approach can succeed if it does not unite the efforts of various engineering disciplines from all over the world under one umbrella to offer a beacon of modern solutions to the global infrastructure. Experts from the various engineering disciplines around the globe will participate in this series, including: Geotechnical, Geological, Geoscience, Petroleum, Structural, Transportation, Bridge, Infrastructure, Energy, Architectural, Chemical and Materials, and other related Engineering disciplines.

More information about this series at http://www.springer.com/series/15140

Hany El-Naggar · Khalid El-Zahaby ·
Hany Shehata
Editors

Innovative Solutions for Soil Structure Interaction

Proceedings of the 3rd GeoMEast International Congress and Exhibition, Egypt 2019 on Sustainable Civil Infrastructures – The Official International Congress of the Soil-Structure Interaction Group in Egypt (SSIGE)

Editors
Hany El-Naggar
Dalhousie University
Dalhousie, NB, Canada

Hany Shehata
Soil-Structure Interaction Group
in Egypt (SSIGE)
Cairo, Egypt

Khalid El-Zahaby
Housing and Building
National Research Center
Giza, Egypt

ISSN 2366-3405 ISSN 2366-3413 (electronic)
Sustainable Civil Infrastructures
ISBN 978-3-030-34251-7 ISBN 978-3-030-34252-4 (eBook)
https://doi.org/10.1007/978-3-030-34252-4

This Springer imprint is published by the registered company Springer Nature Switzerland AG
The registered company address is: Gewerbestrasse 11, 6330 Cham, Switzerland

Contents

Seismic Fragility Evaluation of Retrofitted Low-Rise RC Structures . . . 1
Mohamed Noureldin and Jinkoo Kim

Discussions About Tunnels Accidents Under Squeezing Rock Conditions and Seismic Behavior: Case Study on Bolu Tunnel, Turkey . . . 13
Marcio Avelino de Medeiros and Moisés Antônio Costa Lemos

Hydraulic Response of an Internally Stable Gap-Graded Soil Under Variable Hydraulic Loading: A Coupled DEM-Monte Carlo Approach . . . 25
Sandun M. Dassanayake and Ahmad Mousa

Assessment of Liquefaction Potential Index Using Deterministic and Probabilistic Approaches – A Case Study . . . 34
Graziella Sebaaly and Muhsin Elie Rahhal

Impact Analysis of Soil and Water Conservation Structures-Jalyukt Shivar Abhiyan- A Case Study . . . 47
Ajay Kolekar, Anand B. Tapase, Y. M. Ghugal, and B. A. Konnur

Failure of Overhead Line Equipment (OHLE) Structure Under Hurricane . . . 54
Chayut Ngamkhanong, Sakdirat Kaewunruen, Rui Calçada, and Rodolfo Martin

Use and Comparison of New QA/QC Technologies in a Test Shaft . . . 64
Patrick J. Hannigan and Rozbeh B. Moghaddam

A Case Study on Buckling Stability of Piles in Liquefiable Ground for a Coal-Fired Power Station in Indonesia . . . 88
Muhammad Hamzah Fansuri, Muhsiung Chang, and Rini Kusumawardani

Strategy for Rehabilitation and Strengthening of Dam - A Case Study of Temghar Dam 107
Pravin Kolhe, Sunil Pradakshine, Gunjan Karande-Jadhav, and Anand Tapase

Factors Affecting Lubrication of Pipejacking in Soft Alluvial Deposits 121
Wen-Chieh Cheng and Ge Li

Effect of Soil–Structure Interaction on Free Vibration Characteristics of Antenna Structure 135
Venkata Lakshmi Gullapalli, Neelima Satyam, and G. R. Reddy

Author Index 145

About the Editors

Dr. Hany El-Naggar is an associate professor of Geotechnical Engineering at Dalhousie University with more than 24 years of experience in civil construction, geotechnical and structural engineering and research in Canada and overseas. He has participated in several geotechnical and structural investigations and is experienced in the analysis and design of foundations and soil-structure interaction of buried infrastructure. Dr. El Naggar and his research team has investigated the soil-structure interaction (SSI) effects around buried infrastructure; explored innovative use of tire-derived aggregate (TDA) as a buffer zone to create stress arching and reduce demand on rigid culverts, proposed an earth pressure reduction system using geogrid reinforced platform bridging system to reduce stresses on buried utilities, and developed innovative "cellular" precast concrete pipe system.

Also, he has developed a simplified technique to account for the group Effect in Pile Dynamics; and closed-form solutions for the moments and thrusts in jointed and un-jointed composite tunnel lining systems, designed several foundation systems ranging from machine foundations subjected to dynamic loads to raft foundations for underground structure, as well as several tunnels and underground structures in Canada, Europe, Middle East, and the USA. The findings from Dr. El Naggar's research have been reported in more than 100 technical publications covering both experimental and numerical work in the fields of soil-structure interaction, buried infrastructure, and concrete pipes. He has given several workshops and short courses on various geotechnical engineering topics including Foundation Design, Geotechnical Earthquake Engineering, and Buried Infrastructure Design and they are very well received by practitioners. He is the recipient of the 2016–2017 Outstanding Teaching Award from the Faculty of Engineering, Dalhousie University. In addition, he received the 2005–2006 Outstanding Teaching Award from the Department of Civil and Environmental Engineering at Western. Dr. El Naggar won the 2006 L. G. Soderman Award, the 2005 R. M. Quigley Award, and the 2004 Milos Novak Memorial Award.

Dr. El Naggar is the current chair of the Buried Structures Committee of the Canadian Society of Civil Engineers (CSCE); also, he is the ex-chair the of the New Brunswick chapter of the Canadian Geotechnical Society; he is also member of the technical committee on buried structures of the Canadian Highway Bridge Design Code (CHBDC). He is a member of several professional associations including the Association of Professional Engineers of Nova Scotia; the Association of Professional Engineers of Ontario (PEO); the Canadian Society of Civil Engineering (CSCE), the Canadian Geotechnical Society (CGS); the International Society for Soil Mechanics and Geotechnical Engineering (ISSMGE); and the Engineers Syndicate, Egypt.

Prof. Dr. Eng. Khalid El-Zahaby graduated from the Faculty of Engineering, Cairo University (Egypt), in 1985 with honors. He obtained his M.Sc. from Cairo University in 1989 and his Ph.D. from Civil Engineering Department at North Carolina State University, Raleigh, NC (USA) in 1995. Upon graduation, he joined the Housing and Building National Research Center (HBRC) as a research assistant and was promoted in different positions. He is currently the HBRC Chairman. Dr. El-Zahaby is the recipient of the 2013 ICC Global Award of Atlantic City, NJ (USA) and has been a geotechnical consultant for almost 30 years. He is currently serving as an assessor for geotechnical consultants in Egypt. Dr. El-Zahaby is also the Principal Investigator of a research project funded by the STDF (Egypt) and NSF (USA) on the assessment of seismic hazards on wind turbines.

Hany Shehata is the founder and CEO of the Soil-Structure Interaction Group in Egypt “SSIGE.” He is a partner and vice-president of EHE-Consulting Group in the Middle East, and managing editor of the “Innovative Infrastructure Solutions” journal, published by Springer. He worked in the field of civil engineering early, while studying, with Bechtel Egypt Contracting & PM Company, LLC. His professional experience includes working in culverts, small tunnels, pipe installation, earth reinforcement, soil stabilization, and small bridges. He also has been involved in teaching, research, and consulting. His areas of specialization include static and dynamic soil-structure interactions involving buildings, roads, water structures, retaining walls, earth reinforcement, and bridges, as well as, different disciplines of project management and contract administration. He is the author of an Arabic practical book titled “Practical Solutions for Different Geotechnical Works: The Practical Engineers’ Guidelines.” He is currently working on a new book titled “Soil-Foundation-Superstructure Interaction: Structural Integration.” He is the contributor of more than 50 publications in national and international conferences, and journals. He served as a co-chair of the GeoChina 2016 International Conference in Shandong, China. He serves also as a co-chair and secretary general of the GeoMEast 2017 International Conference in Sharm El-Sheikh, Egypt. He received the Outstanding Reviewer of the ASCE for 2016 as selected by the Editorial Board of International Journal of Geomechanics.

Seismic Fragility Evaluation of Retrofitted Low-Rise RC Structures

Mohamed Noureldin(✉) and Jinkoo Kim

Department of Civil and Architectural Engineering,
Sungkyunkwan University, Suwon, South Korea
m.nour@skku.edu

Abstract. In this research, a seismic fragility evaluation is used to assess the seismic performance of low-rise RC structures retrofitted with self-centering post-tensioned pre-cast concrete (SC-PC) frame with enlarged beam-ends. The seismic performance of the retrofitted structures is verified by constructing fragility curves for the retrofitted and retrofitted structure. Thirty natural time-history earthquake records are generated based on a design response spectrum to construct the seismic fragility curves. Nonlinear time history (NLTH) analysis is used to obtain the median structural demand. Limit states are defined based on seismic guidelines to represent the median structural capacity. The relation between the demand and capacity is modeled as a lognormal cumulative density function in a closed-form equation. The results of the study show that the SC-PC frame is effective in seismic retrofitting and significantly improves the structure seismic fragility for all limit states. In addition, the results show that the most significant improvement is found for the collapse-prevention limit state where severe earthquakes are expected.

1 Introduction

Recent earthquakes highlighted the deficiencies of the existing structures and the need to retrofit many structures designed with old seismic guidelines or built before the application of seismic design codes. To this end, it is imperative to retrofit the existing structure in an effective way and assess this retrofitting by evaluating the seismic behavior by comparing the seismic vulnerability of the structure before and after retrofitting.

Seismic fragility curves are considered a powerful tool for seismic evaluation of existing and new structures. Recently, Remki et al. (2018) investigated the seismic fragility of 60% of the residential building of the city of Algeria, which has been represented by six building classes. Significant damage is found for masonry buildings and Reinforced concrete buildings. Casotto et al. (2015) developed a robust methodology to derive fragility curves for precast RC industrial buildings; the study emphasized the importance of the connection collapse phenomenon on the fragility of the buildings. Salamia et al. (2019) investigated the effect of applying advanced modeling technique using nonlinear fiber beam-column element on a 2-story RC structure for main shock-after-shock earthquake sequences.

H. El-Naggar et al. (Eds.): GeoMEast 2019, SUCI, pp. 1–12, 2020.
https://doi.org/10.1007/978-3-030-34252-4_1

Retrofitting structures with self-centering techniques manifested itself as an effective retrofitting technique for all types of structures (Chancellor et al. 2014). Kurosawa et al. (2019) introduced a precast prestressed concrete frame system with mild press joints for retrofit of existing reinforced concrete frames. He conducted cyclic loading tests on precast prestressed concrete and cast-in-situ reinforced concrete frame sub-assemblies. The results showed that the retrofitted frames have better performance in control cracking compared to the monolithic RC frames. Laboratory seismic testing (e.g., Englekirk 2002; Buchanan et al. 2011) showed the effectiveness of self-centering concrete frames in limit damage to the structure. Recently, smart materials such as shape memory alloy (SMA) have been used with self-centering precast segmental bridge columns (Nikbakht et al. 2015) and showed good performance in nonlinear static and lateral seismic loading.

In the current study, a seismic fragility evaluation will be conducted for an RC structure retrofitted with an exterior SC-PC frame. The SC-PC frame is attached with post-tensioning (PT) unbonded steel strands running parallel to the beams of the existing structure (Fig. 1). The exterior SC-PC frame will have enlarged beam-ends as shown in (Fig. 2).

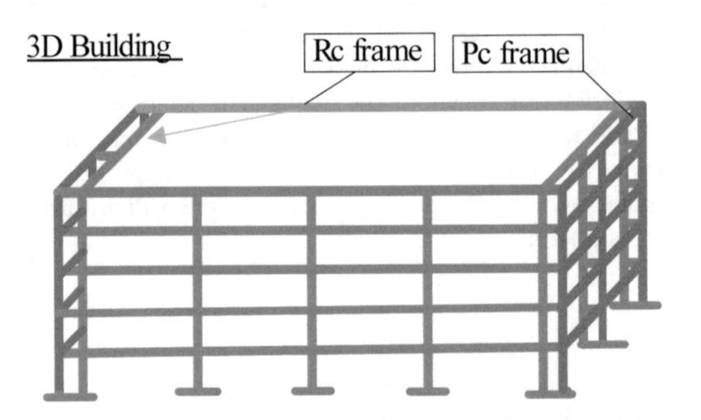

Fig. 1. 3-D RC building retrofitted with self-centering-PC Frame.

2 Mathematical Modeling of the SC-PC Frame

Typical retrofitting PC beams have a rectangular cross-section throughout the length. In this study, the depth of the beam-ends is enlarged as shown in Fig. 2 to increase the capacity of the beam-column connection of the SC-PC frame.

The stress-strain relation of the post-tensioning tendon, (Celik and Sritharan 2004), for Grade 270 prestressing strands, is given in Eq. 1.

$$f_{pt} = \varepsilon_{pt}.E_p.\left[0.02 + 0.98/\left[1 + \left(\frac{\varepsilon_{pt}.E_p}{1.04 \cdot f_{py}}\right)^{8.36}\right]^{1/8.36}\right] \quad (1)$$

where E_p is the elastic modulus of the prestressing steel; ε_{pt} is the strain in the post-tensioning tendon; f_{py} is the yield strength of the post-tensioning tendon.

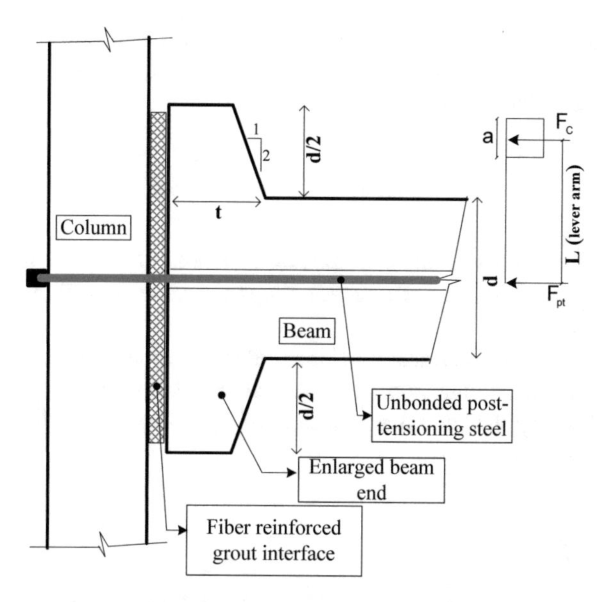

Fig. 2. Details of the beam-column connection of self-centering-PC Frame.

From Fig. 2, the moment capacity of the beam-column connection, M_{cap}, is calculated by multiplying the force developed in the post-tensioning tendon, F_{pt}, with the distance to the resultant concrete compression force, F_c. Based on that, the moment capacity of the beam-column connection is calculated as follows,

$$M_{cap} = F_{pt} \cdot \left(h_g - a\right)/2 \tag{2}$$

where a is the depth of the equivalent rectangular compression stress block corresponding to the compression force, which can be determined using the following equation (ACI-318-14 2014):

$$a = F_c/0.85 f_c^{'} b_g \tag{3}$$

where F_c is the concrete compression force; b_g and h_g are the width and depth of the grout pad at the beam-column interface; $f_c^{'}$ is the unconfined concrete compression strength. At yield of the post-tensioning tendon, M_{cap} can be calculated as

$$M_{cap} = F_{py} \cdot \left(h_g - a\right)/2 \tag{4}$$

When the stress in the extreme concrete compression fiber reaches zero at the beam-end adjacent to the column, the decompression point defines the beginning of a gap opening at the beam-column interface. Taking into consideration the pre-compression introduced by the initial prestressing force, and assuming a linear strain distribution at the critical section, the following equation is used to determine the moment resistance at gap opening, M_{decomp} (Celik and Sritharan 2004).

$$M_{decomp} = f_{pi} \cdot I / \left(\frac{h_g}{2} \right) \tag{5}$$

where f_{pi} is the initial stress in the post-tensioning tendon; I is the moment of inertia of the beam section based on the gross section properties; h_g is the height of the grout pad at the interface.

At the beam-column interface, a bi-linear elastic spring is used to model the gap opening between the column and the beam at the decompression level in the PT tendons. When the applied moment exceeds M_{decomp}, the gap increases and the PT tendons start to elongate.

3 Case Study

The effectiveness of the retrofitting is verified through 2-D frame model. The analytical models for the bare and retrofitting frames are shown in Fig. 3. For the bare frame, beam and column sections are 250 mm × 500 mm and 400 mm × 400 mm, respectively, in all stories. 4D20 and 8D16 longitudinal reinforcement bars with 413 MPa-grade steel are used for beam and column, respectively. D8 @ 150 mm transverse reinforcement is used for both. The compressive strength of the concrete is taken as 21 MPa and cracked sections are assumed for beam and column based on ACI-318R (2014). The model is designed for gravity loads with 7.0 and 2.0 kN/m^2 for dead and live loads, respectively. Plastic hinges are defined for the beam and column section at both ends based ASCE/SEI 41 (2013).

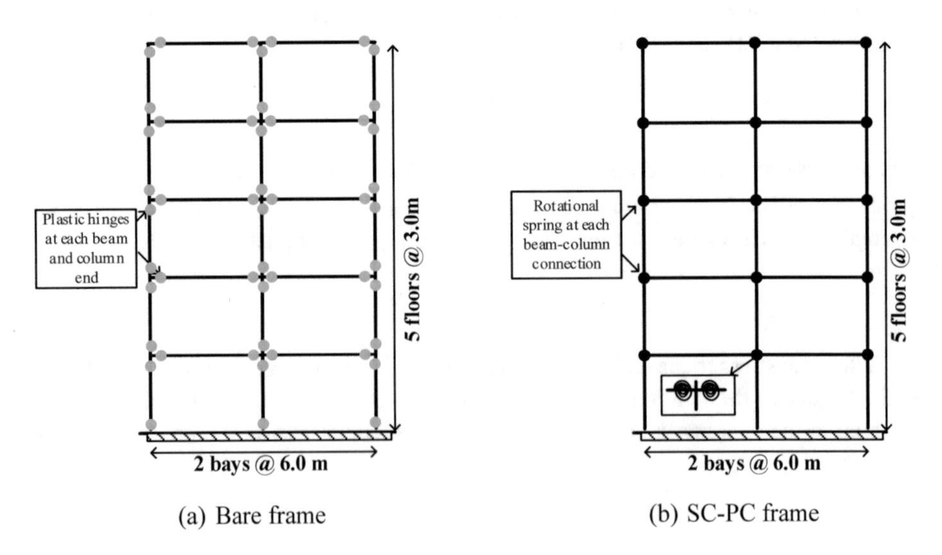

Fig. 3. Bare frame and the SC-PC frame analytical models

The yield strength of the post-tensioning tendons of the SC-PC frame f_{py}, is 1,757 MPa and the initial stress after losses, f_{pi}, is 820 MPa; grout strength is taken as 64.0 MPa and the nominal compressive concrete strength, f_c', is 34.0 MPa. Three tendons with a total area of 380.0 mm^2 are placed in a 300 by 600 mm, concrete section of the PC beam. The connection between the bare frame and the retrofitting frame should be rigidly connected laterally at the connection between the beam and column.

4 Earthquake Records for Seismic Performance Assessment and Fragility Curves

In order to conduct a seismic performance assessment and constructing the fragility curves, a set of earthquake records is required. PEER NGA database (2017) is used to generate natural time history records matching the design spectrum. Figure 4 shows the target (design) response spectrum used for generating the set of the earthquake time-history records. For the construction of the fragility curves, 30 earthquake ground-motion records are obtained from PEER NGA Database (2017) and listed in Table 1.

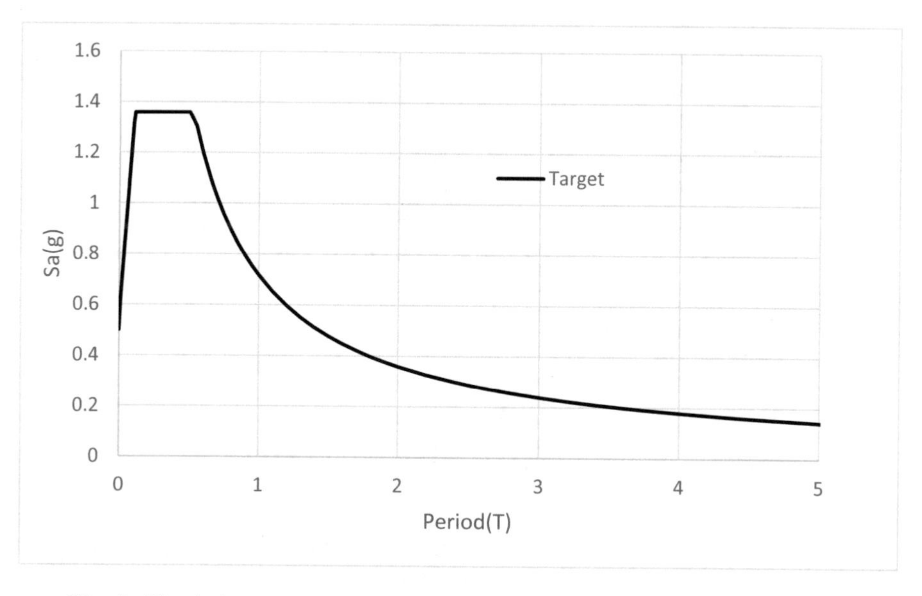

Fig. 4. The design response spectrum used for generating the time history records

Table 1. List of the earthquake records used for constructing fragility curves.

Earthquake name	PEER code	Station
San Fernando	RSN-68	LA-Hollywood Stor FF
Friuli Italy-01	RSN-125	Tolmezzo
Imperial Valley-06	RSN-169	Delta
Imperial Valley-06	RSN-174	El Centro Array #11
Superstition Hills-02	RSN-721	El Centro Imp. Co. Cent
Superstition Hills-02	RSN-725	Poe Road (temp)
Loma Prieta	RSN-752	Capitola
Loma Prieta	RSN-767	Gilroy Array #3
Cape Mendocino	RSN-828	Petrolia
Landers	RSN-848	Coolwater
Landers	RSN-900	Yermo Fire Station
Northridge-01	RSN-953	Beverly Hills - 14145 Mulhol
Northridge-01	RSN-960	Canyon Country - W Lost Cany
Kobe Japan	RSN-1111	Nishi-Akashi
Kobe Japan	RSN-1116	Shin-Osaka
Kocaeli Turkey	RSN-1148	Arcelik
Kocaeli Turkey	RSN-1158	Duzce
Chi Chi Taiwan	RSN-1244	CHY101
Chi Chi Taiwan	RSN-1485	TCU045
Duzce Turkey	RSN-1602	Bolu
Manjil Iran	RSN-1633	Abbar
Hector Mine	RSN-1787	Hector
Kern County	RSN-12	LA-Hollywood Stor FF
El Alamo	RSN-22	El centro Array #9
Parkfield	RSN-30	Cholame-Shandon Array #5
Borrego Mtn	RSN-38	LB-Terminal island
Friuli Italy 01	RSN-121	Barcis
Gazli USSR	RSN-126	Karakyr
Tabas Iran	RSN-138	Boshrooyeh
Trindad	RSN-280	Rio Dell Overpass - FF

SAP2000 (2015) software is used for conducting the nonlinear time history analyses. A material nonlinearity at the ends of the beams and columns (plastic hinges) is modeled using frame element in the software library. Beams and columns are rigidly connected for the un-retrofitted structure. A multi-linear elastic link at the connection between the beam and column of the SC-PC frame is used to model the behavior of the recentering effect of the tendon and the resisting moment at the gap opening.

5 Fragility Curve Formulation

Seismic fragility curves show the probability of a system reaching a limit state as a function of a seismic intensity measure such as spectral acceleration. Seismic fragility is obtained from the results of incremental dynamic analysis, where many nonlinear dynamic analyses are conducted with increasing intensity of the selected time history records. Fragility is described by the conditional probability that the structural capacity C fails to resist the structural demand D. It is generally modeled as a log-normal cumulative density function (Cornell et al. 2002) given by

$$P[C < D|SI = x] = 1 - \Phi\left[\frac{\ln\left(\widehat{C}/\widehat{D}\right)}{\sqrt{\beta_{D/SI}^2 + \beta_C^2 + \beta_M^2}}\right] = 1 - \Phi\left[\frac{\ln\left(\widehat{C}/\widehat{D}\right)}{\beta_{TOT}}\right] \quad (6)$$

where as $\Phi[\cdot]$ is the standard normal probability integral, $\widehat{C}$ is the median structural capacity associated with a limit state, D is the median structural demand, and β_C is system collapse uncertainty, uncertainty in the structural demand $\beta_{D/SI}$ and modeling uncertainties β_M. In this study, the total system collapse uncertainty β_{TOT} is assumed to be 0.6 according to FEMA P695 (2009). In this study, the performance levels such as immediate occupancy (IO), life safety (LS), and collapse prevention (CP) limit states are used based on ASCE/SEI 41 (2013) definition.

6 Seismic Fragility Evaluation

In this section, the seismic fragility of the structure before and after retrofitting are compared and evaluated. Figure 5 shows the fragility curves for the bare structure at three damage states: (a) IO, (b) LS, and (c) CP, which corresponds to maximum inter-story drift (MIDR) of 1%, 2%, and 3%, respectively. As can be seen, at the 50% probability of reaching or exceeding the limit state, the spectral values are 0.33, 0.53, and 0.7 g, respectively. In addition, the sensitivity of the seismic fragility probability values to the change in the spectral acceleration is insignificant after the latter one reaches 1.0 g.

Figure 6 shows the fragility curves for the retrofitted structure at three damage states. At the 50% probability of reaching or exceeding the limit state, the spectral values are 0.40, 0.67, and 0.84 g, respectively, for IO, LS, and CP limit states. In addition, the sensitivity of the seismic fragility probability values to the change in the spectral acceleration is insignificant after the latter one reaches 1.45 g. This means that the spectral acceleration at 50% probability of exceeding the limit state is increased by 21.1%, 26.4%, and 20.0% for the retrofitted structure compared to the un-retrofitted case for the IO, LS, and CP limit states, respectively.

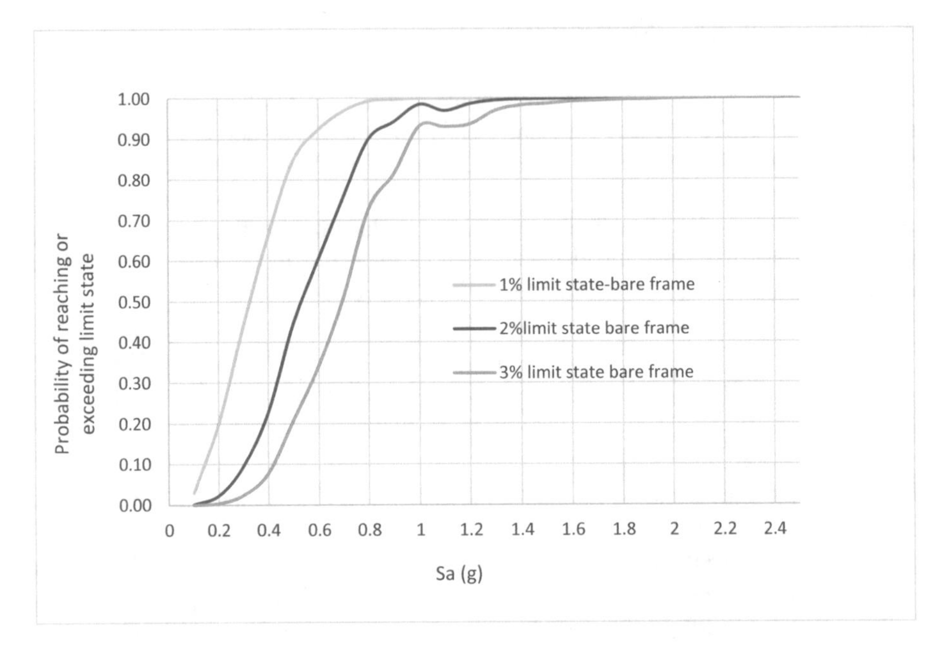

Fig. 5. Seismic fragility curves for the bare frame (un-retrofitted structure)

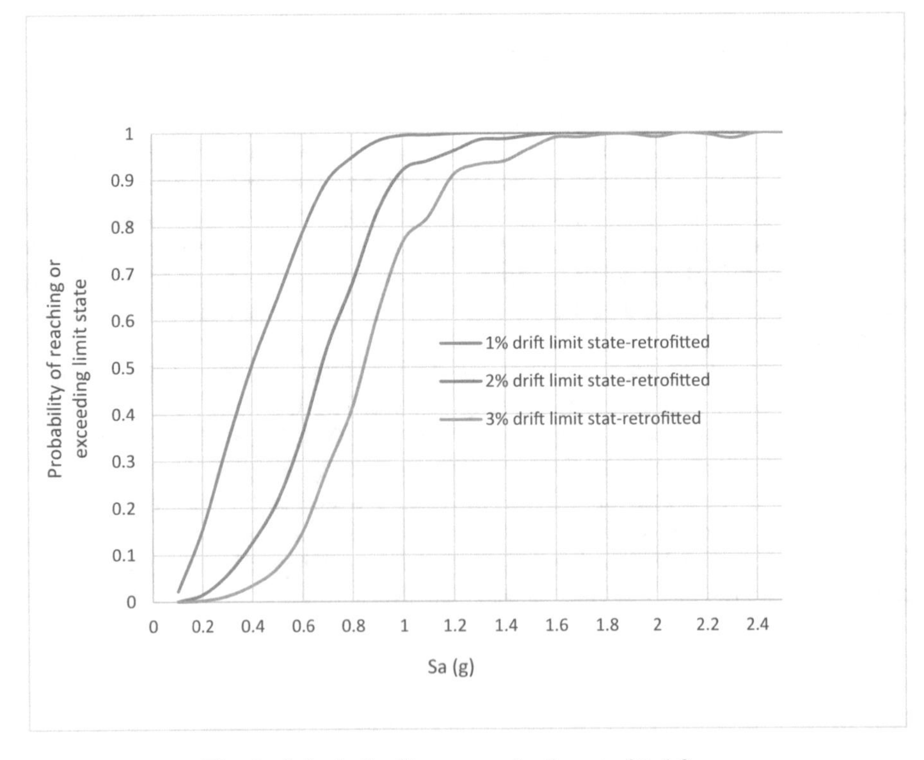

Fig. 6. Seismic fragility curves for the retrofitted frame

Figure 7 shows the fragility curves for the immediate occupancy (IO) limit state for the bare and retrofitted structures. The figure shows that the range of the spectral acceleration where there is a significant improvement in seismic fragility lies between 0.3 g to 0.8 g. The spectral acceleration associated with the fundamental period of the structure is around 0.6 g. This means that the retrofitting scheme can improve the seismic fragility for the IO limit state by decreasing the probability of reaching or exceeding the IO limit state from 92% to 80%. This means that the total percent improvement in seismic fragility is 15% of the original.

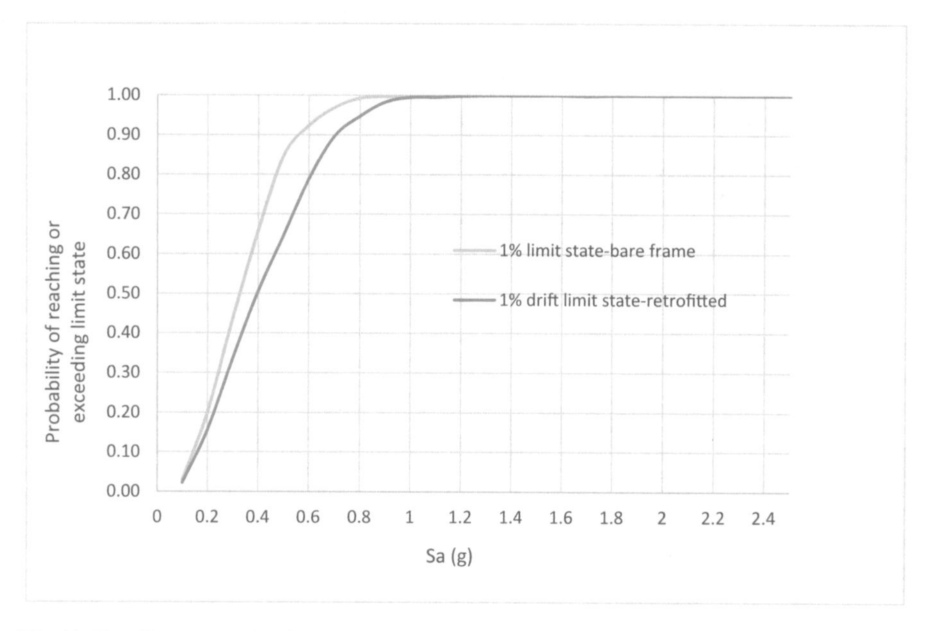

Fig. 7. Fragility curves for the immediate occupancy (IO) limit state for the bare and retrofitted structures.

Figure 8 shows the fragility curves for the immediate occupancy (LS) limit state for the bare and retrofitted structures. The figure shows that the range of the spectral acceleration where there is a significant improvement in seismic fragility lies between 0.35 g to 1.1 g. At 0.6 g (i.e. the spectral acceleration of the structure under investigation), the probability of reaching the LS limit state decreased from 60% to 37%. This means that the total percent improvement in seismic fragility is 62.1% of the original.

Figure 9 shows the fragility curves for the immediate occupancy (CP) limit state for the bare and retrofitted structures. As it can be seen, at 0.6 g (i.e. the spectral acceleration of the structure), the probability of reaching the CP limit state decreased from 35% to 16%. This means that the total percent improvement in seismic fragility is 118% of the original.

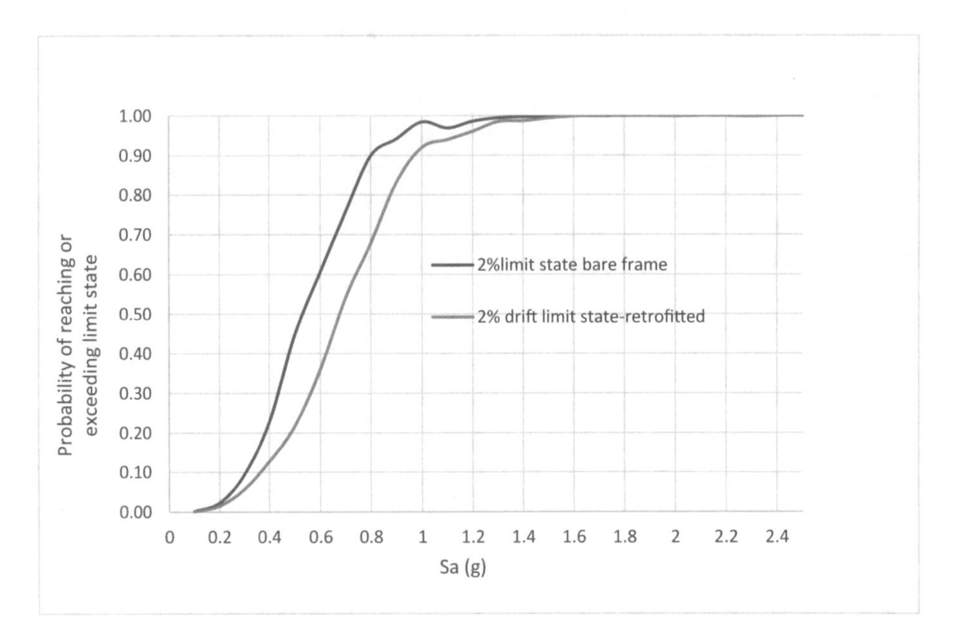

Fig. 8. Fragility curves for the immediate occupancy (LS) limit state for the bare and retrofitted structures.

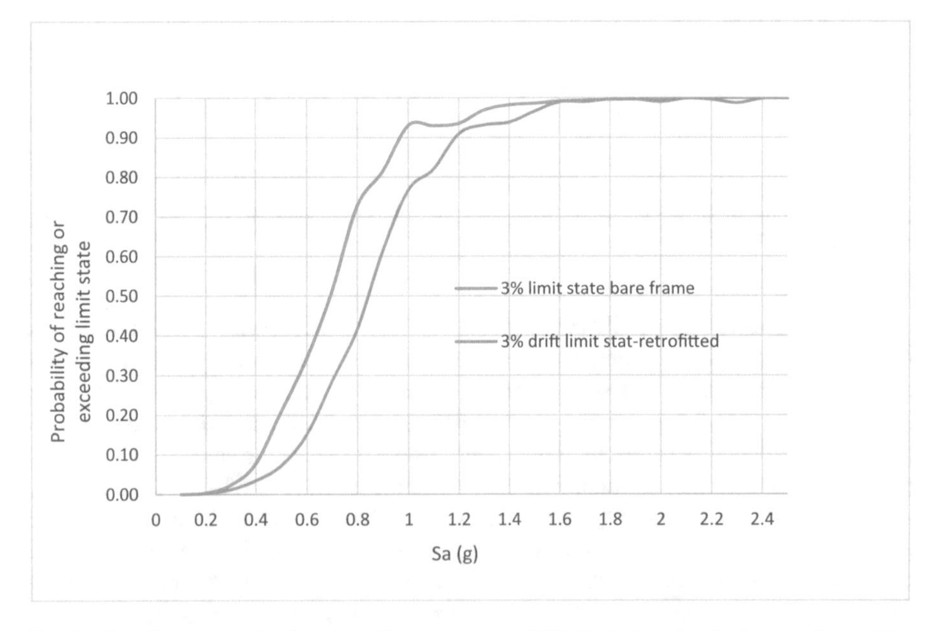

Fig. 9. Fragility curves for the immediate occupancy (CP) limit state for the bare and retrofitted structures.

7 Conclusions

In this study, the seismic behavior of a 5-story RC frame building retrofitted with SC-PC frame externally is assessed through fragility analysis. The retrofitted structures showed that SC-PC frame is effective in improving the seismic behavior by increasing the spectral acceleration at 50% probability of reaching or exceeding the limit states IO, LS, and CP by 21.1%, 26.4%, and 20.0%, respectively, compared to the un-retrofitted case. In addition, at the spectral acceleration associated with the natural period of the structure, the retrofitting scheme can improve the seismic fragility for the IO limit state by 15% of the un-retrofitted. However, for the (LS) limit state, the improvement reached 62.1% of the un-retrofitted. In the case of the CP limit state, the seismic fragility is improved by 118%. This indicates that, for the low-rise RC building investigated in the current study and from the seismic fragility standpoint, the SC-PC retrofitting technique is effective in improving the seismic fragility of all limit states (IO, LS, CP). The significant improvement found in the case of severe limit states such as collapse prevention limit state suggests that this retrofitting technique is very effective for high seismic zones.

Acknowledgments. This research was supported by the "Basic Science Research Program" through the National Research Foundation of Korea (NRF) funded by the Ministry of Education (NRF-2017R1D1A1B03032809).

References

ACI 318R-14 (ACI): Building Code Requirements for Structural Concrete (ACI 318-14) and Commentary. American Concrete Institute, Michigan, USA (2014)

ASCE/SEI 41-13: Seismic Evaluation and Retrofit of Existing Buildings. American Society of Civil Engineers, USA (2013)

Buchanan, A.H., Bull, D., Dhakal, R., MacRae, G., Palermo, A., Pampanin, S.: Base Isolation and Damage-Resistant Technologies for Improved Seismic Performance of Buildings, Technical Report 2011-02 for the Royal Commission of Inquiry into Building Failure Caused by the Canterbury Earthquakes. University of Canterbury, Christchurch, New Zealand (2011)

Casotto, C., Silva, V., Crowley, H., Nascimbene, R., Pinho, R.: Seismic fragility of Italian RC precast industrial structures. Eng. Struct. **94**, 122–136 (2015). https://doi.org/10.1016/j.engstruct.2015.02.034

Celik, O., Sritharan, S.: An Evaluation of Seismic Design Guidelines Proposed for Precast Concrete Hybrid Frame Systems, ISU-ERI-Ames Report ERI-04425, Submitted to the Precast/Prestressed Concrete Manufacturers Association of California, Final report, Iowa State University of Science and Technology (2004)

Chancellor, N.B., Matthew, R.E., David, A.R., Akbaş, T.: Self-centering seismic lateral force resisting systems: high performance structures for the city of tomorrow. Buildings **4**(3), 520–548 (2014). Open Access Journal, ISSN 2075-5309, published online by MDPI

Cornell, C.A., Jalayer, F., Hamburger, R.O., Foutch, D.A.: Probabilistic basis for the 2000 SAC Federal Emergency Management Agency steel moment frame guidelines. J. Struct. Eng. **128** (4), 526–533 (2002)

Englekirk, R.E.: Design-construction of the paramount—a 39-story precast prestressed concrete apartment building. PCI J. **47**, 56–71 (2002)

FEMA P695: Quantification of Building Seismic Performance Factors. Prepared by the Applied Technology Council for the Federal Emergency Management Agency, Washington, D.C. (2009)

Salamia, M.R., Kashanib, M.M., Goda, K.: Influence of advanced structural modeling technique, mainshock-aftershock sequences, and ground-motion types on seismic fragility of low-rise RC structures. Soil Dyn. Earthq. Eng. **117**, 263–279 (2019). https://doi.org/10.1016/j.soildyn.2018.10.036

Remki, M., Kibboua, A., Benouar, D., Kehila, F.: Seismic fragility evaluation of existing RC frame and URM buildings in Algeria. Int. J. Civ. Eng. **16**, 845–856 (2018). https://doi.org/10.1007/s40999-017-0222-7

Nikbakht, E., Rashid, K., Hejazi, F., Osman, S.A.: Application of shape memory alloy bars in self-centering precast segmental columns as seismic resistance. Struct. Infrastruct. Eng. **11**(3), 297–309 (2015)

PEER, Pacific Earthquake Engineering Research Center: Strong motion database, Berkeley, Calif., US (2017). https://peer.berkeley.edu/peer-strong-ground-motion-databases

Kurosawa, R., Sakata, H., Qu, Z., Suyama, T.: Precast prestressed concrete frames for seismically retrofitting existing RC frames. Eng. Struct. **184**, 345–354 (2019). https://doi.org/10.1016/j.engstruct.2019.01.110

Sap2000, ver 18: Analysis Reference Manual. Computer and Structures, Berkeley, USA (2015)

Discussions About Tunnels Accidents Under Squeezing Rock Conditions and Seismic Behavior: Case Study on Bolu Tunnel, Turkey

Marcio Avelino de Medeiros(✉) and Moisés Antônio Costa Lemos

University of Brasilia, Brasilia, Brazil
eng.marcioavelino@gmail.com, moisesaclemos@gmail.com

Abstract. Tunnel Engineering has been modernizing to increase workplace safety and minimize the risk of collapse, mainly in the construction phase, which, statistically, has the highest number of accidents. This accidents on tunnels can be caused by different origins, depending on factors such as geology, natural disasters, poor planned or executed project and even lack of maintenance. However, in tunnel construction, two cases draw the attention of engineering due to the damage caused: effects of seismic behavior and squeezing rock. The impact of seismicity must be analyzed in a tunnel project, since it is an underground work, whose mass is directly susceptible to this action. Besides, the combination of unfavorable effects on the safety of the work with the effects of earthquakes must be verified. The impacts of squeezing should also be taken into account. Thus, rocks studies that present such behavior must be made for a better understanding of their behavior and critical deformations due to the loading imposed by the excavation. This paper discusses the lessons learned in tunnel constructions in regions where the squeezing rock phenomenon occurs and its complication in seismic zones. Therefore, a case study was carried out as a result of the accident in the Bolu tunnel, Turkey, whose results demonstrated the rupture of the twin tunnels due to an earthquake and the presence of squeezing rock.

Keywords: Tunnels · Squeezing rock · Seismic behavior · Accident · Bolu

1 Introduction

Engineering constructions have become increasingly involved concerning the technologies employed and studies carried out. In geotechnical engineering, these complexities are visible and are mainly due to the attempt to avoid unpredicted behavior of soils and rocks and also to impacts caused by factors such as exceptional loads and anthropic or natural actions, such as earthquakes.

Tunnels are one of the examples where the difficulty in predicting the behavior of the construction becomes clear. The forecasts for loading, stresses, strains and the unpredictability of the geomechanical model generate more considerable attention by researchers. In these works, great attention is needed for factors such as the quality of the soil and/or rock and the damages caused by earthquakes. Due to the effects and difficulty to predict the behavior of the system, it is possible to highlight as problems that deserve much attention: squeezing in rocks and seismicity.

H. El-Naggar et al. (Eds.): GeoMEast 2019, SUCI, pp. 13–24, 2020.
https://doi.org/10.1007/978-3-030-34252-4_2

The problems of squeezing rocks is a topic of study always addressed by experts (Sharma 1985; Barla 2001; Singh et al. 2007) due to strains caused over time. One of the causes of more significant strain in tunnels is associated with the plastification of the intact rock due to the redistribution of stresses due to the excavation of the massif, surpassing its resistance. According to Aydan et al. (1993), if the deformation occurs instantaneously and explosively, it is called rock-bursting. However, if the deformation occurs over time, it is named squeezing rock. Therefore, squeezing will occur when a particular combination of induced stresses and material's properties in compresses zones around the tunnel beyond its shear strength limit (Barla 2001).

In addition, natural factors, such as the earthquake, are always a significant warning in regions susceptible to this action. Earthquake affects various regions of the globe, and tunnel engineers and experts are concerned about their effects before, during and after construction (Kontagianni and Stiros 2003; Yu et al. 2017; Fabozzi et al. 2017). Therefore, it is clear that the structures of a tunnel must support not only the geostatic stresses but also accommodate additional strain imposed by movements of earthquakes without fail. For many tunnels, with some exceptions including submerged tunnels (suspended), it is well established that the worst situation for the structure is caused by shear waves traveling perpendicular to the tunnel axis (Sandoval and Bobet 2017).

Thus, the damage caused by both phenomena can be of high impact. An example in which both the squeezing and the earthquake brought considerable damage was in the Bolu Tunnel in Turkey. The Bolu Tunnel is part of the Transit European Motorway (TEM). During its construction, its structure failed due to the earthquake and squeezing rock, causing delays in the execution, enormous financial losses, and the use of more robust techniques for the stabilization of the massif.

Therefore, it is possible to see that damage caused by tunnels can cause significant impacts on the structure of the work and cause financial losses, death of workers or users of the tunnels during the accident, and even problems of people movement if the tunnel is an essential traffic route.

The present study aims to discuss the effects of seismicity and squeezing rocks in tunnel stability based on a case study in the Bolu tunnel, Turkey, and analyze possible solutions to avoid collapse.

2 The Bolu Tunnel, Turkey

The Bolu tunnel was designed as a double tunnel with three lines at each entrance. The construction started at the Asarsuyu portal in 1993 and the Elmalik port in 1994. The length is 3236 m in the right tunnel, while the left tunnel is 3287 m long (Fig. 1).

The cross-section ranges from approximately 140 to over 220 m^2, depending on the support system. Between the tunnels, there is a rock pillar 40 m wide. The maximum stress in situ in the tunnel is referred to the depth of 250 m (Dalgıç 2002)

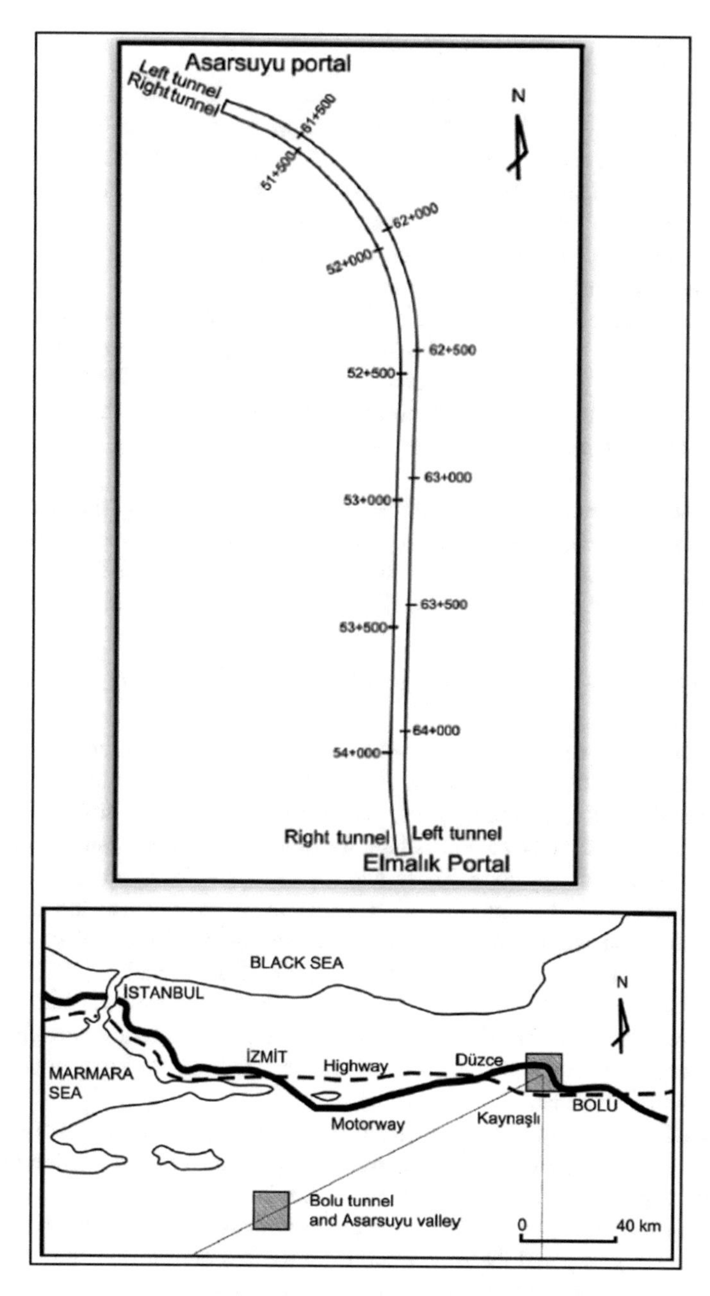

Fig. 1. Bolu Tunnel Localization (Dalgiç 2002)

2.1 Details of the Tunnel Construction

Initially, the tunnel support was designed according to the New Austrian Tunneling Method (NATM) in the 1990s. This type of support project was based on the rock classification, which was following the Austrian standard and was thus defined as a flexible support system (ONORM B 2203). The project provided a typical type of support, which allows a deformation tolerance of 25 cm.

During the first phase of the project, only the effect of the North Anatolian Fault (NAF) zone was considered while the main problems were due to the contact planes, which had no correspondence with the NAF. This caused the sparse classification of the rock as the standard NORM B 2203 and led to the project embargo in the year 1995.

In the following year (1996), in another stage of the project ONORM B 2203 was modified. However, the deformations were not controlled because the flexible supports were used, and as a consequence, there was the collapse of the tunnel on the Elmalik side in 1997.

After the collapse, the deformations were controlled using the abutment drift method in 1997 and the Bernold method, which are more robust support systems. Thus, it is possible to notice that the Bolu tunnel project has been revised several times, but the particularity is due to the tunnel collapse due to the Düzce earthquake.

2.2 Geological and Geomechanical Conditions of the Region

2.2.1 Geology

The Bolu tunnel is located in the North-Anatolian fault zone, an active seismicity zone. Geology along the alignment of the Bolu tunnel is of the Yedigöller Formation, composed mainly of amphibolite, gneiss, meta-diorite and quartz-diorite. Overlapping this unit is the formation Ikizoluk (metasedimentary), composed basically of phyllites, schists, quartzite, and limestone. These two formations make up the Asarsuyu area.

The Elmalik zone is composed of an intrusion of granites and the Findicak formation, which consists of sandy limestone, limestone sandstone, clay. In Fig. 2 is showed the geological profile of the route along the Bolu tunnel.

In the Asarsuyu section, the presence of Amphibolite is observed, followed by two bands of limestone interspersed by fractured marbles and sandstones. Already in the Elmalik section, the geology is composed of granite, sandstone and claystone intrusions along the tunnel alignment.

The results of X-Ray Diffractometry (XRD) showed that in both zones, the acaricidal smectite is the predominant clay, while chlorite and a mixed fraction of chlorite-vermiculite appear in small proportions. That can be seen in Fig. 3.

Note that the most present clay minerals in both zones are the smectite from the Montmorillonite group and the Muscovite from the group of micas, the Elmalik zone being much more abundant in smectite than the Asarsuyu zone.

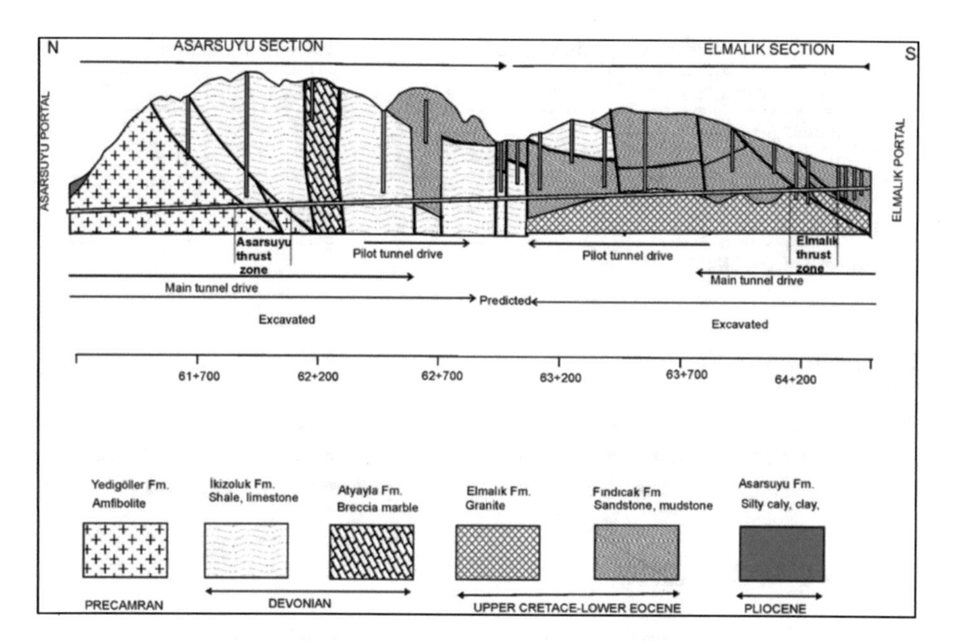

Fig. 2. Geological profile throughout the tunnel (Dalgiç 2002)

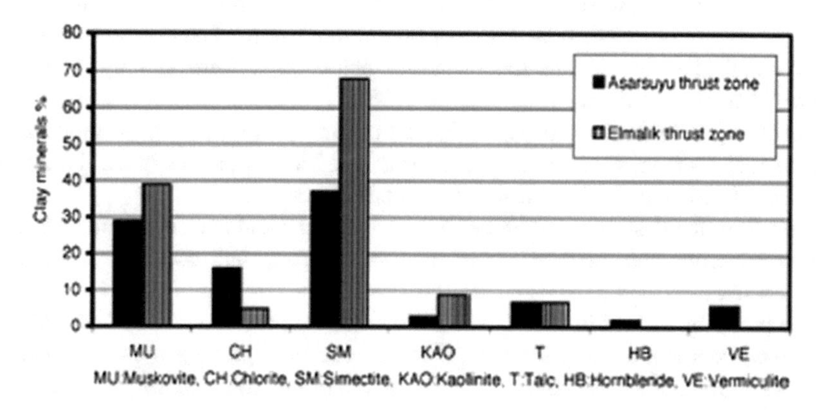

Fig. 3. Minerals in fault zones on long tunnel alignment (Dalgiç 2002)

2.2.2 Geomechanical Model – Parameters

The geomechanical model of the zones of impulse had its parameters obtained from consolidated and drained direct shear test of the clays present in the faults. Both peak and residual resistance parameters were obtained. The values obtained are shown in Table 1.

Table 1. Resistance parameters of impulse zones throughout the tunnel

Source	Impulse Zone	Nº Teste	φ_{peak} (°)	c_{peak} (kPa)	$\varphi_{residual}$ (°)	$c_{residual}$ (kPa)
Yedigöller	Asarsuyu	2	28	25	24	0
İkizoluk		1	25	100	31	0
Findicak	Elmalik	9	18	130	12	90
Elmalik		2	37,5	18	10	16

Source: Dalgiç (2002).

3 Methodology

This research was mainly based on literature review on causes and measures taken to mitigate the effects of the Bolu tunnel collapse in Turkey and on the discussions of other authors about the effects of seismicity and squeezing rock in tunnels.

4 Effects of Seismicity

The idea that tunnels are poorly vulnerable to the effects of seismicity is at first illusory. In areas where there are no flaws or active tectonism, this may even be true, as in the European Northwest. However, in tunnels constructed in regions where there are active faults, seismicity is an essential cause of failure, both in the construction phase (e.g., Bolu Tunnel, Turkey) and in the operating phase (e.g., Kern County Tunnel and Wrights Tunnel) (Kontogianni and Stiros 2003).

According to Jamarillo (2017), the impact of earthquakes on tunnels can usually be grouped into two: mass rupture or mass tremors. The rupture includes liquefaction, slopes instability and displacements of the faults or fractures.

Therefore, the design of the tunnel support system should not only consider a limit state considering the earthquake as a static action, but consider its dynamic characteristic.

The study by Jamarillo (2017) shows that the support dimensioned for the static conditions according to the SEM-NATM is well optimized, but for dynamic conditions due to a system, it does not have enough support capacity to guarantee the equilibrium, safely, of the structure.

5 Effects of Squeezing Rock

The squeezing effects is a time-dependent convergence behavior of tunnel excavation duration time (Barla 2001). When a tunnel is excavated, occurs stresses redistribution around the surface of the massif. The tangential stresses around the tunnel walls become large and exceed the uniaxial compressive strength of the rock mass in the tangential direction of an analyzed point, and the plastification occurs around the tunnel. The rock mass then fail, and the ruptured zone propagates slowly in the radial direction (Singh et al. 2007).

The squeezing phenomenon is associated with rocks of poor quality, that is, large deformations and low resistance properties. According to Barla (2001), based on experience, there are some complex rocks where squeezing can occur, if the loading conditions are sufficient to initiate this behavior, for example, gneiss, mica-schist, calcite tectonic zones or faults), shales.

The discontinuities orientation, such as bedding and schistosity, play a significant role in the appearance and development of large deformations around the tunnels and, therefore, also in the behavior of squeezing.

The pore pressure and the piezometric load are factors that influence the stress-deformation behavior of the rock. Therefore, the rock worsening its quality will tend to occur "squeezing" because its strength has decreased.

The techniques adopted in the excavation and tunnel support work, i.e., the excavation sequences and the number of excavation steps adopted, including the stabilization measures adopted, may influence the general conditions of excavation stability and, consequently, the deformations ("squeezing").

5.1 Squeezing Rock Shapes

In addition to the causes that may influence tunnel construction to have a "squeezing" it is possible to identify the forms in which it occurs. According to Aydan (1989), it is possible to identify three types of failure form (Fig. 4) caused by the closure as a consequence of squeezing: (a) complete shear failure, (b) buckling failure and (c) shearing and sliding failure.

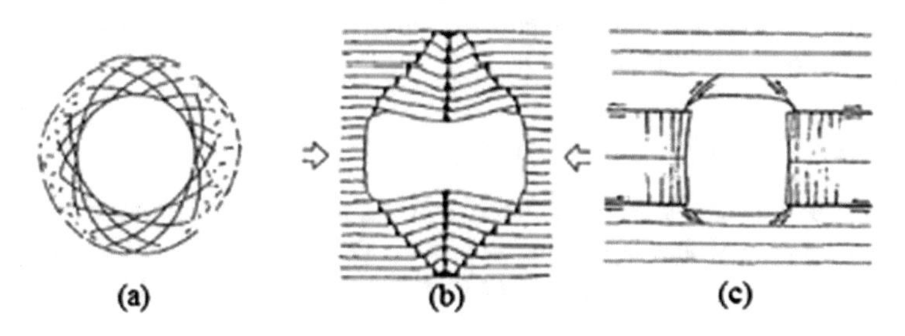

Fig. 4. Squeezing rock shapes (Aydan 1989)

Complete shear failure involves the complete shear process of the medium in comparison with rock-bursting, in which the initiation of the shear process is followed by separation and sudden surrounding rock displacement.

The buckling failure is generally observed in metamorphic rocks (phyllite, mica schist) or thin-layer seismic rocks (limestone, shale, silt, sand, marine evaporites).

Shear and slip failure is observed in sedimentary rocks with thick bedding and involves sliding along layers of bedding and shear of intact rock.

5.2 Squeezing Rock Prediction

In the literature, on tunnel constructions it is possible to identify simple ways of estimating the squeezing phenomenon in rocks. Among the various forms, empirical and semi-empirical methods most be used for predicting the behavior of rocks.

Among the empirical methods, we can mention the Singh et al. (1992) and the method of Goel et al. (1995). For semi-empirical methods, we have Jethwa et al. (1984), Aydan et al. (1993), Hoek and Marinos (2000).

6 Accident Occurred in 1999

6.1 Bolu Tunnel Failure

The magnitude of the Düzce earthquake in 1999 in the Bolu region was 7.2 on the Richter scale, with a maximum horizontal acceleration measured during the earthquake of 80% of the gravitational acceleration (0.81•g).

Although the magnitude of the earthquake was worse than an earlier event (Kocaeli earthquake, 7.4 on the Richter scale), which produced an acceleration of between 25% and 35% of gravitational, its epicenter was 150 km from the tunnel, already the Düzce earthquake had an epicenter just 20 km from the west portal of the tunnel. Thus, although the magnitude of the previous earthquake was more significant, what commanded the effect of seismicity was the epicenter of the post-earthquake.

The Düzce earthquake associated with the North Anatolian fault zone caused the collapse of a clayey area of about 300 m from the portals, which was temporarily supported by 25 cm of projected concrete and 6 to 9 m long rock anchor bolts (Kontogianni and Stiros 2003).

6.2 Solutions Adopted in Bolu Tunnel

In the construction of the Bolu Tunnel, two main types of support structure were adopted in the area where squeezing occurred: flexible support and robust support. According to Darlgiç (2002), the uses of both in the construction of the tunnel can be evaluated as follows:

The principle that a flexible concrete layer should act as a thin crust is not valid for squeezing failure zones such as those found in the Bolu tunnel. More robust reinforcements for tunnel stability should be used in these types of work.

According to the principle of flexible support (Fig. 5), the reinforcement of the massif should be achieved with projected concrete, risers, steel meshes, and anchors. This principle affirms that maximum deformation occurs in areas where deformation cracks are placed. However, with the placement of these cracks, there was an increase in the thickness of the plastification zone. Measurements taken from the Elmalik side indicated that the plastic zone extended to the surface. Moreover, the quality of the mass is negatively affected by the alteration proved by the tie rods.

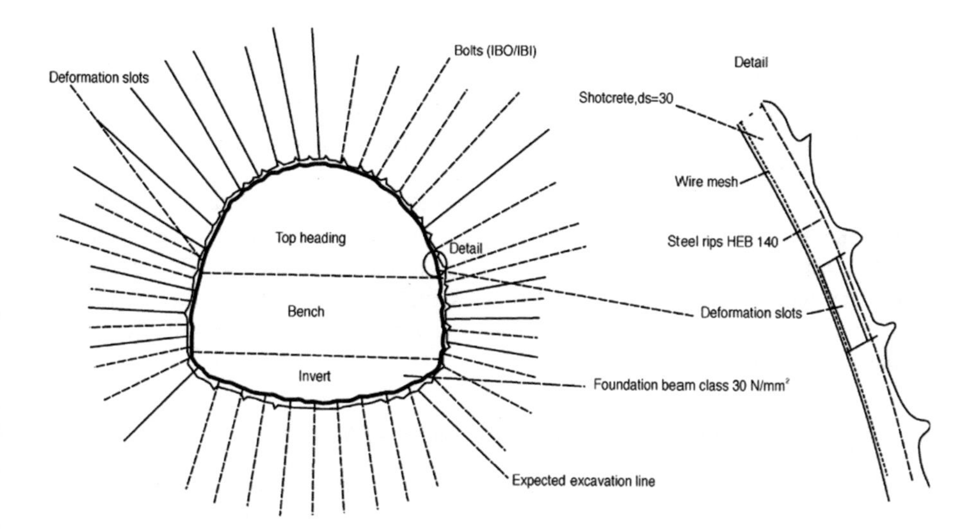

Fig. 5. Minimum shotcrete (Geoconsult 1996)

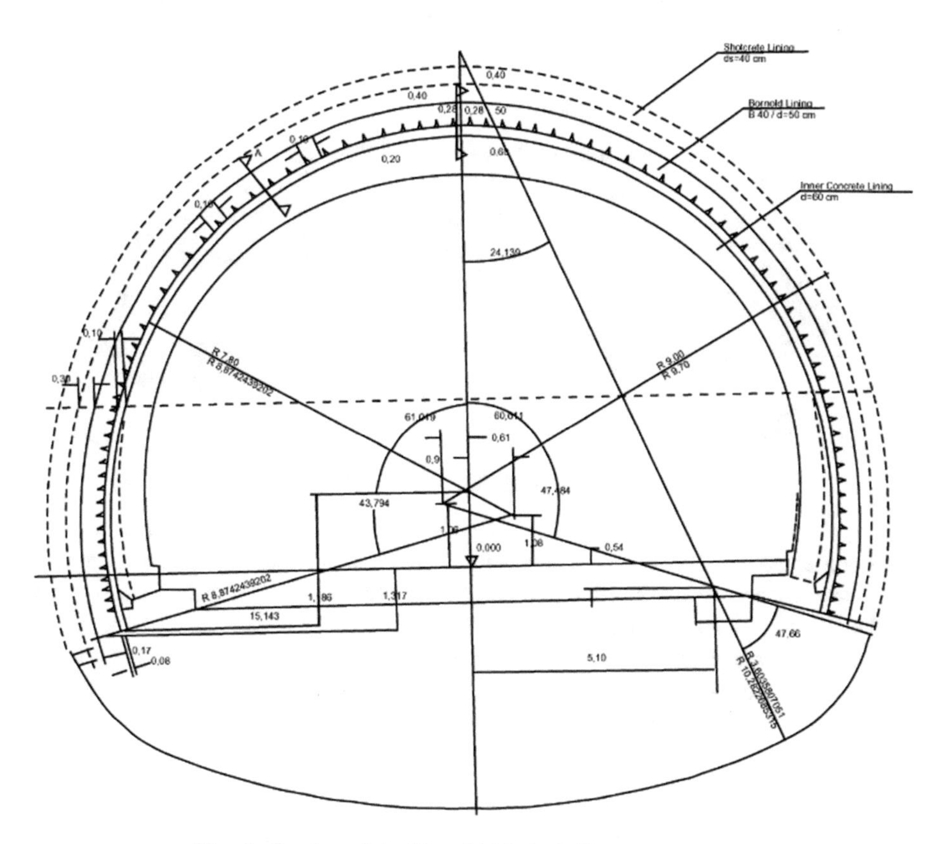

Fig. 6. Section of the Bernold Method (Geoconsult 1996)

It is evident that the acceptable deformation rate in the fault zone and the other in the low-quality area must be greater than 2 mm per month when installing the last layer of the support. In this case, a part of the deformation must be supported by the final coat. Therefore, a reinforced final layer should be required under squeezing mass conditions.

The Bernold support (Fig. 6) was successfully applied in sedimentary rock conditions and in a small fault zone where the angle of internal friction of the soil is higher than 15°, while the abutment drift method (Fig. 7) was successfully applied in a central fault zone, which contains an internal friction angle of less than or equal to 15°. By application of these methods under adverse soil conditions, the deformations were limited to the normal range.

6.3 Similar Accidents and Construction Technics Adopted

The San Francisco Wrights Railroad Tunnel (currently decommissioned) suffered a 7.7 magnitude earthquake on the Richter scale during operation (1906) that caused the collapse of about 100 m along the San Andreas fault (Prentice and Ponti 1997 *apud* Kontogianni and Stiros 2003). The earthquake caused the formation of two surfaces of parallel ruptures, with only one of them provoking the instability of the tunnel.

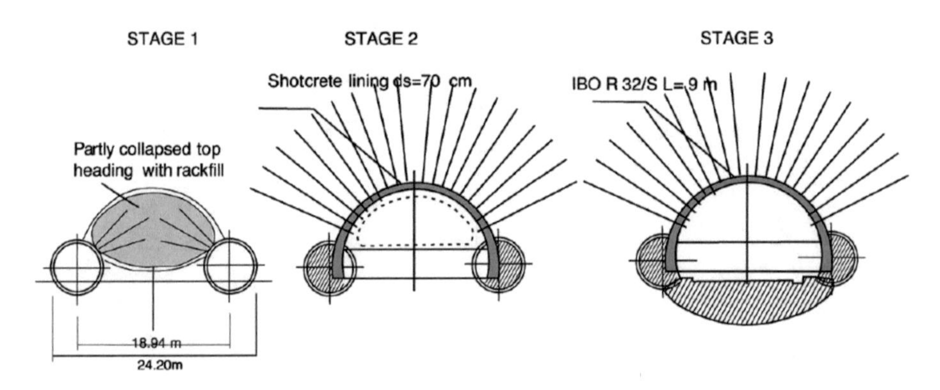

Fig. 7. Transversal section of abutment drift method (Geoconsult 1996)

Another earthquake-related accident in a tunnel was the Kern Country Tunnel in the 1953 Kerny Country Earthquake of 7.5 magnitudes on the Richter scale (SCDECDC 2002 apud Kontogianni and Stiros 2003). It was associated with Wolf's failure White, which caused a lateral compressive displacement component responsible for folds in the interior and portal parts of the tunnel and the rails, which were pushed through the 46 cm thickness support.

7 Conclusions

A right prediction of seismicity by adopting higher return times is essential for a good tunneling project since the consequences of high seismic activity can be catastrophic, causing the tunnel to collapse, as was the case with Bolu. Also, special attention should be paid to earthquakes during the construction phase in tunnels under squeezing rock, whose closure may lead to very sharp deformations amplified by the effect of the earthquake. For this reason, the dimensioning of the support should not be made considering only the static conditions of construction and operation, but also the dynamic conditions due to an earthquake, also analyzing the possible related epicenters.

Acknowledgments. Department of Geotechnics, University of Brasilia (UnB), Coordination for the Improvement of Higher Education Personnel (CAPES) and National Council for Scientific and Technological Development (CNPq) for the supporting.

References

Aydan, O., Akagi, T., Kawamoto, T.: The squeezing potential of rock around tunnels: theory and prediction. Rock Mech. Rock Eng. **2**, 137–163 (1993)

Barla, G.: Tunnelling under squeezing rock conditions. Tunnelling mechanics. In: Kolymbas, D. (eds.) Eurosummer-School in Tunnel Mechanics, Innsbruck, pp. 169–268. Logos Verlag, Berlin (2001)

Darlgiç, S.: Tunneling in squeezing rock, the Bolu tunnel, Anatolian Motorway, Turkey. Eng. Geol. **67**, 73–96 (2002)

Fabozzi, S., Licata, V., Autuori, S., Bilotta, E., Russo, G., Silvestri, F.: Prediction of the seismic behavior of an underground railway station and a tunnel in Napoli (Italy). Undergr. Space. **2**, 88–185 (2017)

Geoconsult: Bolu tunnel redesign technical report, Antolian Motorway, Gümüsova – Gerede, report no: 45.110/R/2118, KGM, Ankara (1996). 10 pp

Goel, R.K., Jethwa, J.L., Paithakan, A.G.: Tunnelling through the young Himalayas – a case history of the Maneri-Uttarkashi power tunnel. Eng. Geol. **39**, 31–44 (1995)

Hoek, E., Marinos, P.: Predicting tunnel squeezing problems in weak heterogeneous rock masses. Tunn. Tunn. Int. **32**, 45–51 (2000). pp. 45–51: part one; pp. 33–36: part two

Jethwa, J.L., Singh, B., Singh, B.: Estimation of ultimate rock pressure for tunnel linings under squeezing rock conditions – a new approach. In: Brown, E.T., Hudson, J.A. (eds.) Design and Performance of Underground Excavations, ISRM Symposium, Cambridge, pp. 231–238 (1984)

Kontagianni, V.A., Stiros, S.C.: Earthquakes and seismic faulting: effects on tunnels. Turkish J. Earth Sci. **12**, 153–156 (2003)

Sandoval, E., Bobet, A.: Effect of frequency and flexibility ratio on the seismic response of deep tunnels. Undergr. Space **2**, 125–133 (2017)

Sharma, V.M.: Prediction of closure and rock loads for tunnels in squeezing ground. Ph.D. thesis, IIT, New Delhi, India (1985)

Singh, B., Jethwa, J.L., Dube, A.K., Singh, B.: Correlation between observed support pressure and rock mass quality. Tunn. Undergr. Space Technol. **7**, 59–74 (1992)

Singh, M., Singh, B., Choudhari, J.: Critical strain and squeezing of rock mass in tunnels. Tunn Undergr. Space Technol. **22**, 343–350 (2007). https://doi.org/10.1016/j.tust.2006.06.005

Yu, H., Yuan, Y., Bobet, A.: Seismic analysis of long tunnels: a review of simplified and unified methods. Underg. Space **2**, 73–87 (2017)

Hydraulic Response of an Internally Stable Gap-Graded Soil Under Variable Hydraulic Loading: A Coupled DEM-Monte Carlo Approach

Sandun M. Dassanayake and Ahmad Mousa(✉)

School of Engineering, Monash University Malaysia, Jalan Lagoon Selatan, 47500 Bandar Sunway, Selangor Darul Ehsan, Malaysia
ahmad.mousa@monash.edu

Abstract. Seepage-induced fines migration, referred to as suffusion, conceivably degrades the hydraulic performance of soil masses subjected to water flow. Suffusion susceptibility of the soil (i.e. internal instability) is empirically evaluated using the grain size distribution (GSD) with little emphasis on the permeation process of fines (i.e. mobile fines) in the matrix. Upon experiencing different hydraulic gradient histories, the mobile fines can detach (unclog) or filtrate (clog) in the pore space of the load bearing soil skeleton (i.e. primary fabric). Moreover, clogging and unclogging phenomena are instrumental to temporally altering the hydraulic responses, mainly the hydraulic conductivity (K) and critical hydraulic gradient required to initiate internal erosion of the soil. Therefore, real-time predictions of the hydraulic response of the soil can reveal the initiation and progression of suffusion. This study attempts to forecast the hydraulic response of the soil by quantifying the statistical distribution of K values under mobile fines contents (f_c). To this end, the discrete element method (DEM) has been employed to estimate the void ratio distribution of the primary fabric. The statistical Monte Carlo simulations have been performed using the Kozeny-Carman equation and the void ratio distribution to estimate the statistical distribution of K values. A set of pilot-scale experiments were conducted on internally stable gap-graded soil subjected to stepwise hydraulic loading histories to allow validation for the numerical results. Based on the measured pore pressure distribution parallel to the seepage path, the possible clogging and filtration of fines were inferred, and the K variations were estimated. The change in GSD of the soil after testing was used as a measure for the mobile fines fraction. An average of 6–8% mass loss (from the fine fraction) was observed over a 45-min duration of the controlled-head flows. The experimentally observed K ($1–3 \times 10^{-4}$ m/s) and f_c ranges fall well within the non-parametric bounds of the predicted K and f_c (<10%) ranges.

Keywords: Temporal suffusion · DEM · Statistical techniques · Fines migration · Inverse modeling

H. El-Naggar et al. (Eds.): GeoMEast 2019, SUCI, pp. 25–33, 2020.
https://doi.org/10.1007/978-3-030-34252-4_3

1 Introduction

Seepage induced fines migration that leads to internal erosion in water retaining earth structures has been the prime cause of nearly half of the reported earth dam failures (Fell et al. 2003). The initial phase of this process, known as suffusion or internal instability, induces changes to the grain size distribution (GSD), void ratio (e) and hydraulic conductivity (K) of the soil without altering its physical structure (i.e. volume) (Fannin and Slangen 2014). Seepage flow initiates the suffusion process by transporting the "loose" fines that do not contribute to the load transfer (i.e. mobile fines fraction, f_c), through the interstitial pore network of the load bearing primary fabric (i.e. coarser fraction) in the soil matrix. It is hypothesized that the mobile fines reside in the pore space of the primary fabric. Yet, the complexity and the concealed nature of the suffusion phenomenon renders quantifying the value of f_c extremely challenging. As such, reliable prediction of the subsequent changes to the GSD, e and K values is unattainable.

Empirical studies show that gap-graded soils typically experience suffusion due to the high presence of fines (Langroudi et al. 2013). However, with excessive fines content (i.e. more than 35%), these soils demonstrate internal stability (Chang and Zhang 2013). Nonetheless, a mediocre local migration of fines (i.e. a change in K) in internally stable soils is inevitable upon the application of a hydraulic gradient (Indraratna et al. 2018; Marot et al. 2016). The f_c subject to migration within the soil matrix during the suffusion process is regulated by the filtration phenomenon (i.e. clogging and unclogging) (Bianchi et al. 2018). Moreover, the hydraulic loading history experienced by the soil governs the filtration and migration phenomena (Rasheed et al. 2018). However, the reported experiments in the literature suggest that the mobile fines content for an internally stable gap-graded soil is always lower than 10% (Chang and Zhang 2013). Since the maximum f_c has a maximum known limit ($<10\%$), the internally stable soils provide a good reference to quantify the effects of mobile fines on the variation of suffusion affected parameters, such as K of the soil (Staudt et al. 2017).

This study establishes a statistically significant distribution of K values for an internally stable soil. This statistical distribution shows the range of K values for the soil that possesses different f_c values at the initiation of suffusion for distinct hydraulic loadings. To this end, discrete element method (DEM) has been employed to develop the statistical distribution of the primary fabric void ratio (Dassanayake et al. 2019). The numerical simulations are coupled with statistical Monte Carlo simulations (Calamak et al. 2017) to predict the distribution of the K values using the Kozeny-Carman equation (Chapuis 2012). One-dimensional flow tests have been performed to simulate different seepage histories on a gap-graded internally stable soil for validating the numerical results. The statistically predicted K distribution ($K_{pred.}$) has been employed to inversely predict the mobile fines content of soil in real-time using the experimentally observed range of K ($K_{obs.}$) values.

2 Methodology

The study comprises both experimental and numerical approaches. A modified permeameter was employed to apply variable stepwise hydraulic loading on a gap graded soil with fine gravel and sand of alluvial origin (Fig. 1a and b). The soil was assessed to be internally stable according to Chang and Zhang's (2013) method of assessment. A statistical distribution of the hydraulic conductivity subjected to mobile fines fraction was developed by coupling discrete element method (DEM) with the Monte Carlo simulations (Fig. 2).

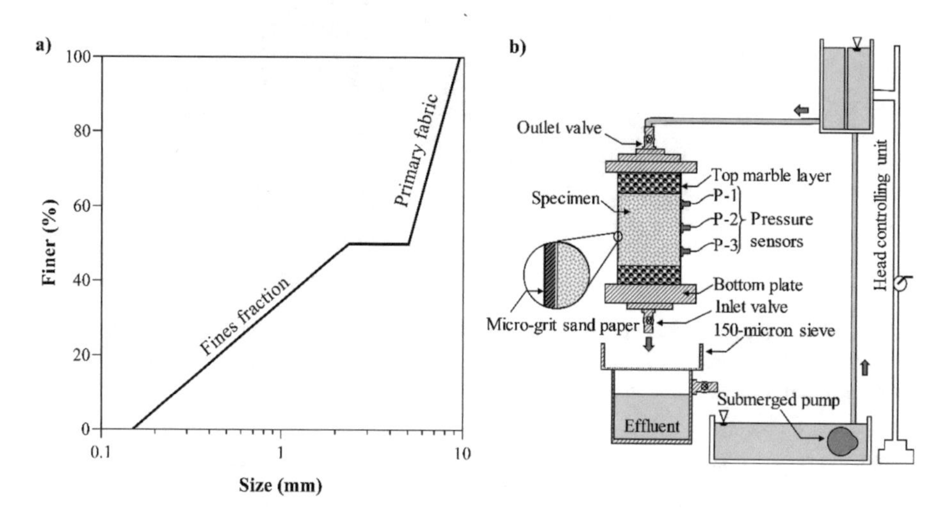

Fig. 1 Experimental approach: (a) grain size distribution of the tested soil; (b) one-dimensional head-controlled flow set-up.

2.1 Flow Tests

A custom-made rigid wall permeameter cell was used to perform the flow tests. It has a diameter of 100 mm, and it can accommodate a sample height up to 400 mm. The cell was attached to a bottom section that has a removable stainless-steel sieve plate with an opening size of 2 mm on which the soil specimen resided during the flow test. Macro grit-sized sandpapers were attached to the rigid permeameter wall to minimize the wall-flow effect. Three pressure transducers were attached to the cell wall at 50 mm intervals to measure the pore water pressure variations across the sample during the test. The collected readings were subsequently used to estimate the global hydraulic gradient variation with time. The effluent collection unit comprises a graduated cylinder and an effluent tank to collect the migrating fines. A 150-μ sieve was used to retain the fines that may escape with the seepage flow.

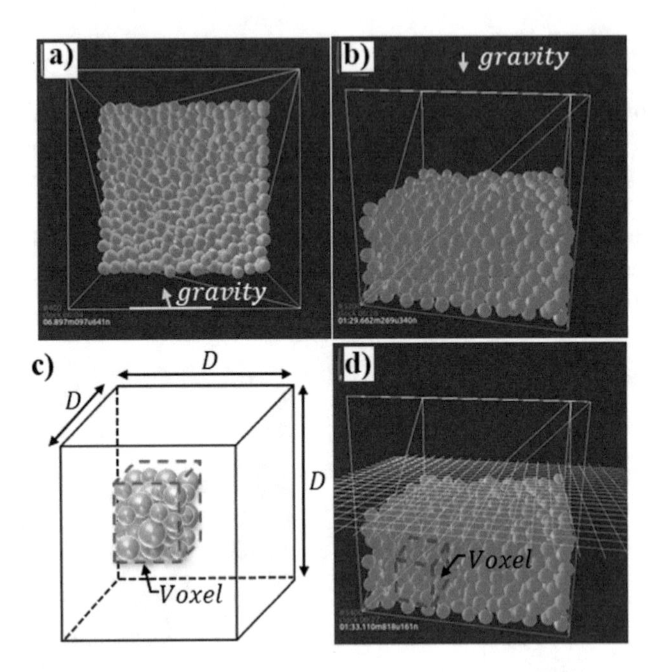

Fig. 2 DEM sampling (Dassanayake et al. 2019): (a) top view of the particle cloud; (b) gravity deposition; (c) schematic of the representative voxel volume; (d) estimating the void ratio distribution after numerical compaction

The soil was compacted in 5 layers on the sieve plate until a 180 mm sample height was achieved. The void ratio of the specimen for all of the experiments was approximately 0.34. The soil's maximum and minimum void ratios were 0.76 and 0.32, respectively. Several preliminary tests warranted the uniformity of the test specimen in term of compaction and GSD. Marble layers were placed on the top and bottom of the sample. These layers were placed to evenly disperse the flow vectors, which allows flow diffusion through the soil medium. Both layers were 50 mm high. The total dry mass of these layers was kept at 1 kg for all the tests, producing a constant overburden stress of 1.3 kPa.

Flow tests were conducted using stepwise controlled head hydraulic loading scenarios. These scenarios included three patterns: S1, S2 and S3 (as given in Fig. 3a). Each hydraulic gradient step applied to the soil has a duration of 15 min. Several preliminary tests, which are not reported herein, showed that a 15 min duration is long enough to completely wash-out the mobile fines in a 50 mm height soil sample (i.e. a representative section). The $K_{obs.}$ was calculated using Darcy equation. After the test, the sample was separated into three representative sections: top, middle, and bottom. The GSDs of these sections were evaluated after the flow (suffusion) test to assess the respective fines losses under the applied hydraulic loading scenarios.

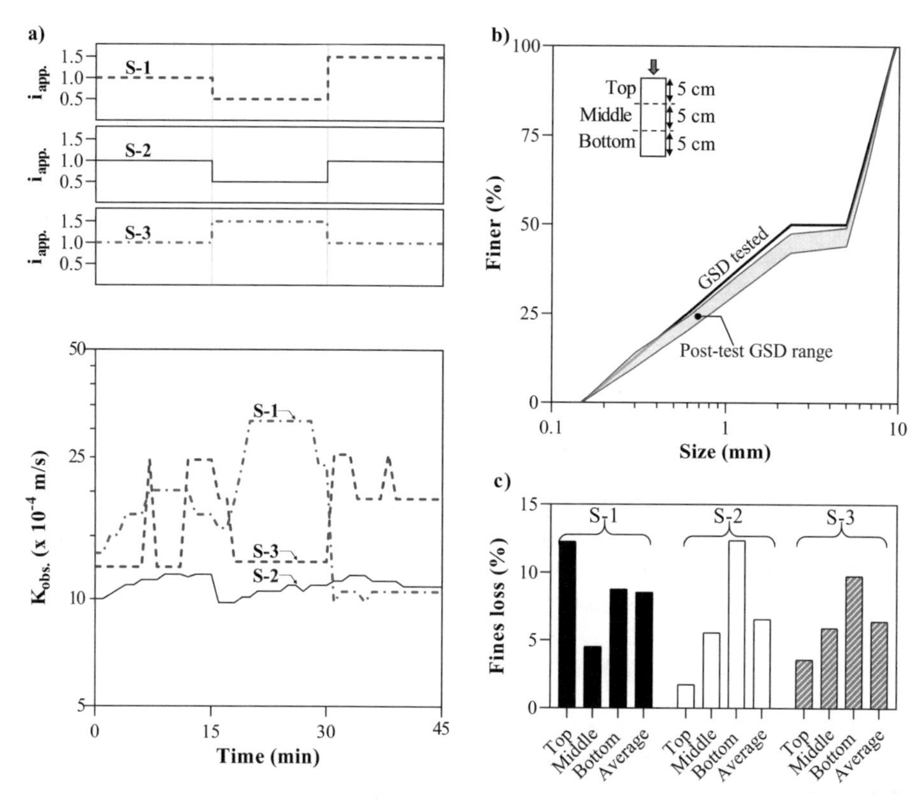

Fig. 3 Hydraulic response of the soil: (a) variation of the experimentally obtained K under variable hydraulic gradient histories; (b) range of the post-test grain size distributions of the soil after the flow tests; (c) fines fraction loss in different sections of the specimen.

2.2 DEM-Monte Carlo Approach

The opensource software code YADE was employed in this study to simulate the particle packing in primary fabric to estimate the statistical distribution of the primary fabric void ratio (e_p). To this end, the particle cloud was generated using 10,000 non-cohesive spherical particles that follow the GSD of the coarse fraction of the tested soil. Then the particle cloud was allowed to freely fall under gravity into a bounding box with the dimensions of 20 × 20 × 20 mm (Fig. 2a–d). After the particle assembly was settled, a 1.3 kPa compaction was applied on to the packing to simulate the test conditions. The void ratio of a given location was estimated using a representative voxel volume (15 × 15 × 15 mm) (Fig. 2c and d). This procedure was repeated for random locations within the bounding box, until a statistically significant distribution of void ratios (approximately hundred e_p values) were obtained. This methodology is further discussed by Dassanayake et al. (2019).

The void ratio of the soil mixture (e) and the fines percentage (f_c) governs the value of e_p (e.g. Chang et al. 2015; Yang et al. 2015) (Eq. 1). If f_c reaches zero (i.e. no fines), e approaches e_p.

$$e_p = [e + (1 - b)f_c]/[1 - (1 - b)f_c] \tag{1}$$

where b is the fraction of the fines content that contribute to the primary skeleton. The value of b ranges from 0 to 1. Equation 1 can be re-expressed, for the case of known e and e_p, to estimate the content of the mobile fines:

$$(1 - b)f_c = (e - e_p)/(1 + e_p) \tag{2}$$

Since f_c is a fractional value (0 to 1), Eq. 2 is lower bounded at 0 and upper bounded at 1. For practical purposes, a 50% of fines (i.e. the upper bound, $f_c = 0.5$) can be considered as the maximum fines content. Therefore, the statistical distribution of the mobile fines content can be estimated for a known e value.

The K of non-cohesive soils can be estimated using Kozeny-Carmen analytical relationship (Eq. 3), which was developed after considering a porous material as an assembly of capillary tubes using the Navier–Stokes equation.

$$K = \left[Cg/\left(\mu_w \rho_w S^2 G_S^2\right)\right] \times \left[e^3/(1+e)\right];\ \text{where } e = e_p - \left(1+e_p\right)(1-b)f_c \tag{3}$$

where S (m^2/kg of solids) is the specific surface of the grain assembly, C is a dimensionless factor to account for the shape and tortuosity of flow channels, g is the gravitational constant (9.81 m/s^2), μ_w is the dynamic viscosity of water, ρ_w is the density of water, G_S is the specific gravity of soil. From Eqs. 1 and 3, the K of the soil assembly can be represented in terms of mobile fines content and the primary fabric void ratio:

$$K = \left[Cg/\left(\mu_w \rho_w S^2 G_S^2\right)\right] \times \left[\left(e_p - \left(1+e_p\right)(1-b)f_c\right)^3/\left(1+e_p - \left(1+e_p\right)(1-b)f_c\right)\right] \tag{4}$$

The frequency distributions were fitted to two distinctive non-parametric probability distributions. These fitted distributions were employed in the Monte Carlo statistical simulation. In this technique, the probability value was randomly generated using a random number generator, available in MATLAB (2019). The random probability value was then employed to inversely sample the e_p and (1–b)f_c values from each probability distribution. The process was simulated for 1,000 different combinations of the parametric values needed to calculate the K value using Eq. 4. The estimated statistical pool was used to develop the upper and lower bounds of the potential K values, given the mobile fines content.

3 Results

The variation of the $K_{obs.}$ values has a narrow range (1×10^{-3} to 3×10^{-3} m/s) (Fig. 3a). This justifies the internal stability of the tested soil. However, the different hydraulic gradient histories have affected the mobile fines fraction. The mobile fines ranged from 2% to 8%. The range of this distribution (6%) is shown in Fig. 3b. The downward flow direction has resulted in a higher loss of fines from the bottom section for all the specimens (Fig. 3c).

Coupled DEM-statistical simulations show that the lower mixture void ratios increase the probability of lower mobile fines contents for the gap-graded soils (Fig. 4a). For a well-compacted gap-graded soil (say e = 0.3), there is a probability of 0.5 to have a mobile fines content in the range of 10% to 20%. When the soil is loosely compacted (say e = 0.7), the probability of having mobile fines content lower than 20% is negligible; however, it is vital to note that there is a probability of 0.5 for the loosely compacted soil to possesses a mobile fines content in the range of 20% to 40%. This counter-intuitively supports the notion that the fines in a well-compacted soil join the load transfer (and they are stationary), whereas loosely compacted soil has more "transportable" fines. Figure 4b shows the variability of the hydraulic conductivities when the mobile fines content increases. The optimum fines content lies approximately around 30% for the soil matrix. However, when this value is increased, the hydraulic conductivities are expected to show two ranges of fines contents. This perhaps indicates the ability of a soil matrix to re-arrange its fore-chain structure to have dubious mobile fines contents. However, further pieces of evidence are needed to support this argument.

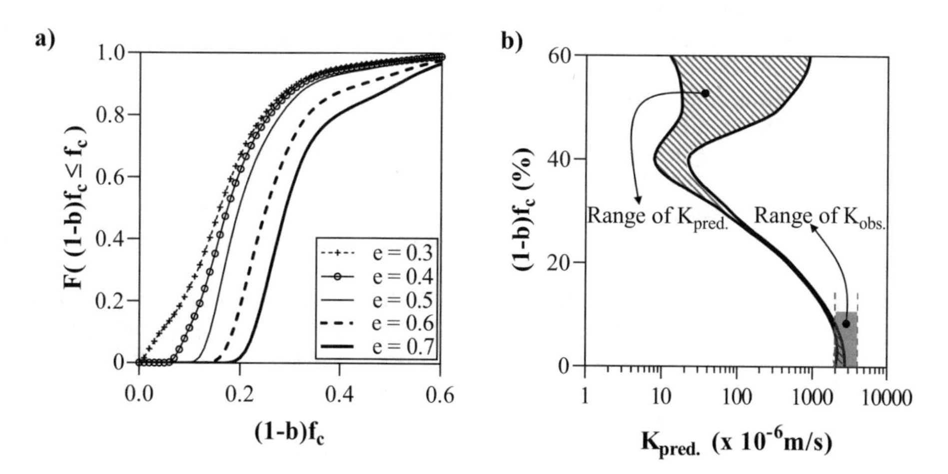

Fig. 4 Statistical distributions: (a) cumulative probability distribution of the mobile fines content for different mixture void ratios estimated from the coupled DEM-statistical approach; (b) potential range for hydraulic conductivities under different mobile fines contents.

4 Conclusion

This study employs a coupled DEM and Monte Carlo simulations technique to develop a statistical distribution for the hydraulic conductivities of an internally stable gap-graded soil. One dimensional downward flow tests were conducted on a internally stable gap-graded soil to obtain a distribution of the hydraulic conductivities, experimentally. The tests simulate three distinct hydraulic loading scenarios. The experimentally obtained Darcy's hydraulic conductivities ranged from 3×10^{-3} to 1×10^{-3} m/s. This range inversely predicts the presence of up to 10% of the mobile fines content in the soil mix. The post-test GSDs show that 2 to 8% of the fines content was washed off with the flow. The close agreement between the predicted and observed mobile fines contents supports the potential use of this technique to real-time assessment of the mobile fines fraction of the soil for a given hydraulic conductivity range.

The future work shall consider extrapolating the statistical distribution to account for different soil types. A generalized model manifested through this notion can warrant the use of probabilistic techniques, such as Bayesian approach (Dassanayake and Mousa 2018), to predict the progression of suffusion phenomenon in geo-applications.

References

Bianchi, F., Wittel, F.K., Thielmann, M., Trtik, P., Herrmann, H.J.: Tomographic study of internal erosion of particle flows in porous media. Transp. Porous Media **122**(1), 169–184 (2018)

Calamak, M., Melih Yanmaz, A., Kentel, E.: Probabilistic evaluation of the effects of uncertainty in transient seepage parameters. J. Geotech. Geoenviron. Eng. **143**(9), 06017009 (2017)

Chang, C.S., Wang, J.Y., Ge, L.: Modeling of minimum void ratio for sand–silt mixtures. Eng. Geol. **196**, 293–304 (2015)

Chang, D.S., Zhang, L.M.: Extended internal stability criteria for soils under seepage. Soils Found. **53**(4), 569–583 (2013)

Chapuis, R.P.: Predicting the saturated hydraulic conductivity of soils: a review. Bull. Eng. Geol. Environ. **71**(3), 401–434 (2012)

Dassanayake, S.M., Mousa, A.: Probabilistic stability evaluation for wildlife-damaged earth dams: a Bayesian approach. Georisk Assess. Manage. Risk Eng. Syst. Geohazards 1–15 (2018)

Dassanayake, S.M., Mousa, A., Kong, D.: Voids distribution of pavement filters under permeating fines: a DEM approach coupled with statistical inference. In: 4th International Conference on Civil Engineering and Materials Science, Bangkok, Thailand (2019)

Fannin, R.J., Slangen, P.: On the distinct phenomena of suffusion and suffosion. Géotechnique Lett. **4**(4), 289–294 (2014)

Fell, R., Wan, C.F., Cyganiewicz, J., Foster, M.: Time for development of internal erosion and piping in embankment dams. J. Geotech. Geoenviron. Eng. **129**(4), 307–314 (2003)

Indraratna, B., Israr, J., Li, M.: Inception of geohydraulic failures in granular soils-an experimental and theoretical treatment (2018)

Langroudi, M.F., Soroush, A., Shourijeh, P.T., Shafipour, R.: Stress transmission in internally unstable gap-graded soils using discrete element modeling. Powder Technol. **247**, 161–171 (2013)

MATLAB (2019). www.mathworks.com

Marot, D., Rochim, A., Nguyen, H.H., Bendahmane, F., Sibille, L.: Assessing the susceptibility of gap-graded soils to internal erosion: proposition of a new experimental methodology. Nat. Hazards **83**(1), 365–388 (2016)

Rasheed, A.K., Dassanayake, S.M., Mousa, A.: Suffusion susceptibility in gap-graded granular soils under variable hydraulic loading. In: Fifteenth International Conference on Structural and Geotechnical Engineering, Cairo (2018)

Staudt, F., Mullarney, J.C., Pilditch, C.A., Huhn, K.: The role of grain-size ratio in the mobility of mixed granular beds. Geomorphology **278**, 314–328 (2017)

Yang, J., Wei, L.M., Dai, B.B.: State variables for silty sands: global void ratio or skeleton void ratio? Soils Found. **55**(1), 99–111 (2015)

Assessment of Liquefaction Potential Index Using Deterministic and Probabilistic Approaches – A Case Study

Graziella Sebaaly(✉) and Muhsin Elie Rahhal(✉)

Faculty of Engineering ESIB, Saint Joseph University of Beirut, Beirut, Lebanon
graziella.sebaaly@net.usj.edu.lb,
muhsin.rahal@usj.edu.lb

Abstract. Based on in situ tests results, the deterministic approaches express the soil liquefaction potential in terms of safety factor and the probabilistic approaches define it in terms of probability of liquefaction. The developed methods for soil liquefaction evaluation predict the behavior of a soil element. However, the liquefaction potential index (LPI) evaluates the performance of the entire soil column and allows a measurement of the severity of liquefaction. In this paper, the liquefaction potential is used in conjunction with deterministic approach and reliability based probabilistic approach using SPT and CPT results. A site located to the north of Beirut where the soil investigation consisted of a total of sixteen boreholes with continuous coring and standard penetration test every 1.5 m of depth and thirty-three boreholes with cone penetration test is subjected to liquefaction potential study. Liquefaction potential index (LPI) is calculated at each borehole location from the obtained safety factor or the calculated probability of liquefaction. Spatial distribution of soil liquefaction potential is presented in form of contours maps of the (LPI) values for each considered method. The results show that the evaluation of LPI using the reliability based probabilistic approach and the CPT results is a more conservative method giving larger area with LPI values higher than 15.

1 Introduction

The soil liquefaction potential is determined based on in situ test results such as standard penetration test (SPT) and cone penetration test (CPT). Several approaches have been used to understand what will happen to a soil element. Deterministic methods have been developed to calculate the safety factor since the original method by Seed and Idriss (1971) including the methods recommended by the NCEER committee (2001), the methods proposed by Boulanger and Idriss (2014). A safety factor value lower than 1 denotes that a soil is liquefiable and a value higher than 1 indicates a non-liquefiable soil.

On the other hand, a probabilistic method has been established by Cetin (2016) and reliability based analyses were undertaken by Juang (2009) and by Sebaaly and Rahhal (2019a, b) to determine the probability of liquefaction by considering the model and parameters uncertainty. The probability of liquefaction gives a description of the level of risk of the liquefaction occurrence for a considered soil element.

H. El-Naggar et al. (Eds.): GeoMEast 2019, SUCI, pp. 34–46, 2020.
https://doi.org/10.1007/978-3-030-34252-4_4

The above methods predict what will happen to a soil element without considering the whole soil profile. However, the liquefaction potential index predicts the performance of the whole column and the consequences of liquefaction on the ground surface.

The liquefaction potential index was first introduced by Iwasaki et al. (1978, 1981, 1982). It predicts the potential of liquefaction to cause damage at the surface level. It was identified that liquefaction risk is very low for LPI = 0, low for 0 < LPI < 5, high for 5 < LPI < 15 and very high for LPI > 15. Other interpretations were proposed for the liquefaction risk based on the values of LPI by Luna and Frost (1998), by Toprak and Holzer (2003), by Sonmez (2003) and by Holzer et al. (2006). The LPI has been also correlated with the different forms of surface effects such as lateral spreading, ground cracking, sand boils and with ground damage near foundations. The LPI combines depth, thickness, and factor of safety against liquefaction (FS) or probability of liquefaction (P_L).

In this paper, the liquefaction potential index is used in conjunction with SPT and CPT deterministic methods and with reliability based probabilistic approach using SPT and CPT results. The liquefaction potential index (LPI) is calculated from the factors of safety (FS) or from the probabilities of liquefaction along the depth at each representative borehole at site located to the North of Beirut. Consequently, the (LPI) values were used for the compilation of liquefaction hazard maps for the considered site.

2 Liquefaction Potential Index

Liquefaction potential index is a single value parameter to evaluate liquefaction potential of a whole soil column. It combines depth, thickness and the indicator of liquefaction occurrence (safety factor or probability of liquefaction). The effects of soil liquefaction are limited to 20 m of depth since no liquefaction damages have been reported for greater depths.

Taking into consideration the above, Iwasaki et al. (1982) proposed the following equation for the liquefaction potential index (LPI):

$$\mathrm{LPI} = \int_{0}^{20} \mathrm{F}\, \mathrm{w(z)dz} \tag{1}$$

Where z is the depth, w(z) is the weighting factor, F is the severity factor and dz is differential increment of depth. The weighting factor gives more importance to the layers closer to the ground surface by following a linear trend and is calculated as follows:

$$\mathrm{w(z)} = 10 - 0.5\mathrm{z} \tag{2}$$

The severity function is a key component. It is related to the safety factor against the initiation of liquefaction but only the soils with safety factors lower than 1. It is given at any depth by the following equations:

$$F = 1 - FS \quad \text{if} \quad FS \leq 1 \tag{3}$$

$$F = 0 \text{ if } FS > 1$$

However, F may be derived from the probability of liquefaction (Juang et al. 2006) by adopting the following definition:

$$F = P_L - 0.35 \text{ if } P_L \geq 0.35 \tag{4}$$

$$F = 0 \text{ if } P_L \leq 0.35$$

The threshold probability of 0.35 was chosen since the likelihood of liquefaction is considered low if P_L is lower than 0.35 and in this case, the liquefaction induced ground failure may be negligible.

The level of liquefaction severity with respect to LPI as per Iwasaki et al. (1982), Luna and Frost (1998), and Microzonation for Earthquake Risk Mitigation (MERM 2003) is given in Table 1.

Table 1. Level of liquefaction severity

LPI	Iwasaki et al. (1982)	Luna and Frost (1998)	MERM (2003)
LPI = 0	Very low	Little to none	None
0 < LPI < 5	Low	Minor	Low
5 < LPI < 15	High	Moderate	Medium
15 < LPI	Very high	Major	High

3 Deterministic Approach

The calculation of the safety factor requires the calculation of the cyclic resistance ratio (CRR) and the cyclic stress ratio (CSR). CRR is a measure of liquefaction resistance and CSR is the representation of seismic loading that causes liquefaction. The safety factor is defined as the ratio of CRR over CSR. In this paper, three methods have been used to calculate the safety factor: the one defined by Youd et al. (2001) using the standard penetration test results, and the methods defined by Idriss and Boulanger (2014); one using the standard penetration test results and one using the cone penetration test results.

For the above cited methods, the cyclic stress ratio is expressed in function of the peak horizontal acceleration at the ground surface generated by the earthquake (a_{max}), the acceleration of gravity (g), the total and effective vertical overburden stresses (σ_{v0} and σ'_{v0}), stress reduction factor (r_d), the magnitude scaling factors (MSF) and the overburden correction factor (K_σ) as follows:

$$CSR_{7.5,\sigma} = \frac{\tau_{av}}{\sigma'_{v0}} = 0.65\left(\frac{a_{max}}{g}\right)\left(\frac{\sigma_{v0}}{\sigma'_{v0}}\right)(r_d)/MSF/K_\sigma \quad (5)$$

However, the two methods differ when estimating the values of r_d, MSF and K_σ. Idriss and Boulanger (2014) derived the equation of r_d using the results of numerous site response analysed and updated the equations of MSF and K_σ in order to account for their variation as a function of the soil characteristics.

In the SPT-based model by Youd et al. (2001), the cyclic resistance ratio CRR is calculated from the Equation below:

$$CRR = \frac{1}{34 - (N_1)_{60cs}} + \frac{(N_1)_{60cs}}{135} + \frac{50}{\left[10(N_1)_{60cs} + 45^2\right]} - \frac{1}{200} \quad (6)$$

where $(N_1)_{60cs}$ = clean sand equivalence of the overburden stress corrected SPT blow count conducted after the correction for fines content.

Idriss & Boulanger (2014) determined the cyclic resistance ratio (CRR) as a function of $(N_1)_{60cs}$ value and as a function of the equivalent clean sand correction of the normalized tip resistance q_{c1Ncs}. These relations are expressed by the following equations:

$$CRR = \exp\left[\frac{(N_1)_{60,cs}}{14.1} + \left(\frac{(N_1)_{60,cs}}{126}\right)^2 - \left(\frac{(N_1)_{60,cs}}{23.6}\right)^3 + \left(\frac{(N_1)_{60,cs}}{25.4}\right)^4 - 2.8\right] \quad (7)$$

$$CRR = \exp\left[\frac{q_{c1Ncs}}{113} + \left(\frac{q_{c1Ncs}}{1000}\right)^2 - \left(\frac{q_{c1Ncs}}{140}\right)^3 + \left(\frac{q_{c1Ncs}}{137}\right)^4 - 2.8\right] \quad (8)$$

For both SPT and CPT results, Idris and Boulanger (2014) expressed the correction for the overburden pressure as a function of the relative density of the soil expressed in terms of $(N_1)_{60cs}$ and q_{c1Ncs} respectively. A new adjustment of the fines content (FC) has been defined.

4 Reliability Based Probabilistic Approach

The evaluation of soil liquefaction is expressed in terms of probability of liquefaction. By adopting a reliability based approach, the probability of liquefaction is calculated by taking into account the parameters uncertainties and the model uncertainty.

The reliability index (β) as defined by Hasofer-Lind is calculated using FORM (first order reliability method) by the means of numerical method (excel spreadsheet). The probability of liquefaction is then determined as $P_L = 1 - \Phi(\beta)$, where Φ is the standard normal cumulative distribution function.

The limit state function for liquefaction triggering may be expressed as: $h(x) = c\ \text{CRR} - \text{CSR}$. x represents the vector of the input parameters required for calculation of CRR and CSR and c is a variable that represents the model state parameter used to describe the model uncertainty.

In this study, the cyclic resistance ratio and the cyclic stress ratio are calculated using the three deterministic methods defined in the previous paragraph. For the methods using the standard penetration test results, the random variables can be limited to $N_{1,60}$, FC, M_w, a_{max}, σ'_v and σ_v. For the methods based on the cone penetration test results, the $N_{1,60}$ is replaced by the q_{c1N}. The random variables and the model parameter are correlated variables assumed to follow a log normal distribution. The uncertainties of the parameters and the model are assessed in terms of mean value and coefficient of variation. The approach considered by FORM is to transform the correlated lognormal variable into normal independent random variable.

Using Cetin database (Cetin 2000), Juang et al. (2009) concluded for the Youd et al. model a mean value for the model parameter of 0.96 and a COV of 0. Sebaaly and Rahhal (2019a, b) characterized the model uncertainty of the Idriss and Boulanger model based on the SPT results by a mean value of 1.03 and a coefficient of variation of zero by using the database reported by Cetin (2016) and the one based on the CPT results by a mean value of 1.33 and a coefficient of variation of zero by employing the database reported by Moss (2003).

For the remaining random variables, the mean value is taken equal to measured value at the considered depth. However, the coefficient of variation is calculated based on the existing data for the soil parameters ($N_{1,60}/q_{c1N}$, FC, σ'_v and σ_v) and is considered based on published data of its typical range for the seismic parameters (M_w and a_{max}).

5 Case Study

The studied area is located on the coast area to the north of Beirut, in an industrial zone. It covers an area of $\sim$79000 m^2. The soil investigation included the execution of sixteen boreholes with continuous coring and standard penetration test every 1.5 m of depth and thirty-three boreholes with cone penetration tests. Eight boreholes have been equipped with PVC standpipes piezometers for water level measurement. Figure 1 shows the location of the executed boreholes within the studied area.

Fig. 1. Boreholes' location plan

5.1 Geological Context of the Site

From a geological perspective, the formation of the considered site is Quaternary formation a (sandy alluvium) overlaying the Cretaceous formation. The Quaternary deposits consist of different soils and material that constitute nonconsolidated layers covering the outcropping rock formations. Within the Beirut area, the different parts of the Quaternary deposits consist, from old to young, of brown reddish soil with cobbles and gravels of limestone origin, yellow sandy soil cover, alluvial quaternary deposits of cobbles and pebbles, sand material called "ramleh" with conglomerates and reddish sandy soil. A groundwater table exists in this area at a shallow depth (Figs. 3 and 4).

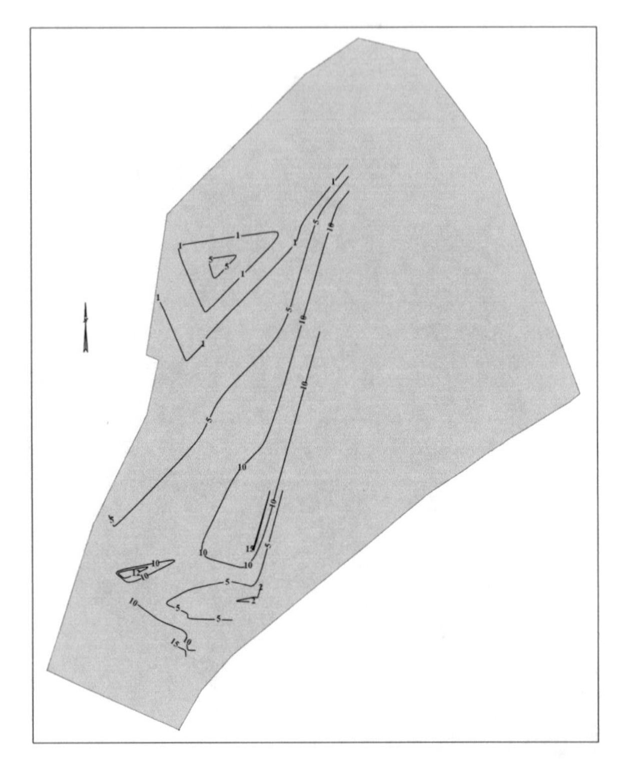

Fig. 2. Contour map of LPI for SPT results using deterministic approach (Youd et al. model)

5.2 Seismic Context of the Region

Lebanon lies across an estimated 1000 km long fault which extends from the seafloor spreading in the Red Sea to the Taurus in Southern Turkey. This fracture system, known as the Levant or Dead Sea fault system, is an extremely important tectonic feature, which accounts for the bulk of seismic activity in the Eastern Mediterranean. For the calculations, a maximum acceleration of 0.25 g and a moment magnitude of 7 are considered according to the Lebanese standards.

5.3 Geotechnical Context of the Site

The obtained core sample are consistent with the geology of the area. The first encountered layer in all the boreholes is a fill material having a variable thickness ranging between 1 m and 2 m. Below the fill material, a sandy layer was encountered having a variable thickness ranging between 3.5 m and 24.5 m. It consists of medium

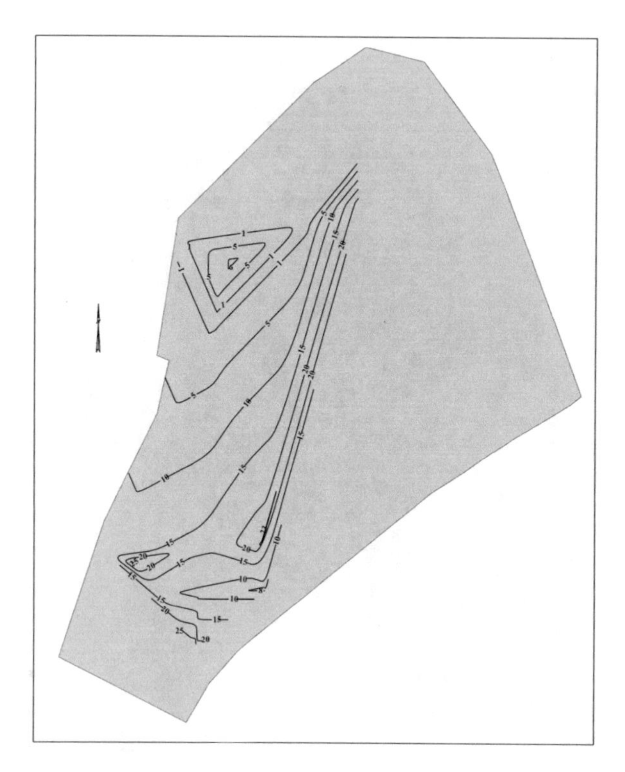

Fig. 3. Contour map of LPI for SPT results using reliability based probabilistic approach (Youd et al. model)

sand, loose to medium dense with variable percentage of fines. The sandy layer overlays a sandy clay/clayey sand layer that itself overlays a very dense sand layer.

The measured water level varies between ~1.2 m and ~1.8 m of depth depending on the location of the boreholes. This water level mainly corresponds to the sea water level.

5.4 Computation of the Liquefaction Potential Index

The liquefaction potential index is calculated for every executed borehole down to 20 m of depth. The evaluation of LPI is done in conjunction with the deterministic methods cited previously for calculation of the safety factor or with conjunction of the reliability based probabilistic approach explained priory to determine the probability of liquefaction.

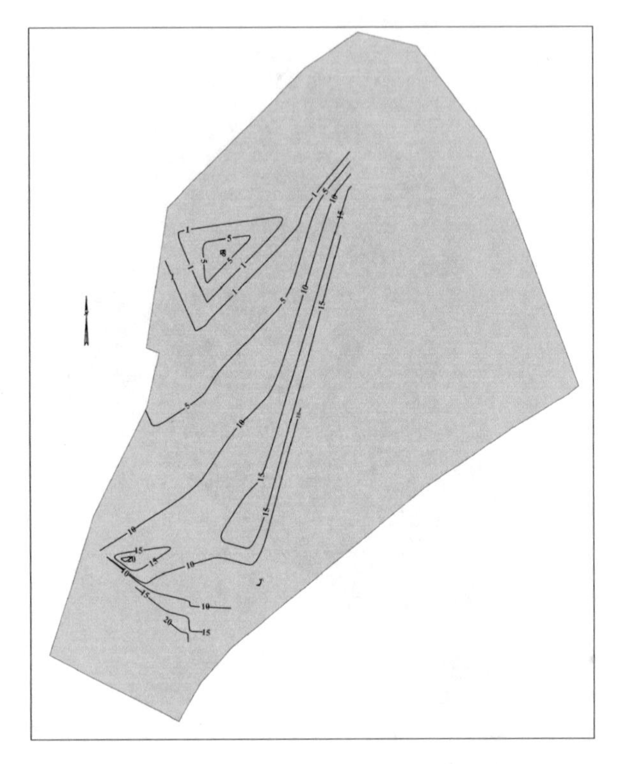

Fig. 4. Contour map of LPI for SPT results using deterministic approach (Idriss and Boulanger model)

Spatial distribution of soil liquefaction potential is presented in form of contours maps of the (I_L) values for each considered method. These LPI contour maps will give an indication of geographic variability of liquefaction effects and different kinds of probable surface manifestations of liquefaction.

5.5 Results and Discussions

Seismic soil liquefaction in terms of LPI is evaluated across the designated site. The contour maps of LPI values are generated showing the spatial distribution of liquefaction potential. Figures 2 through 5 show the results of LPI contour maps using SPT results in conjunction with previously defined deterministic and probabilistic reliability based approach. Based on these figures, the probabilistic methods are more

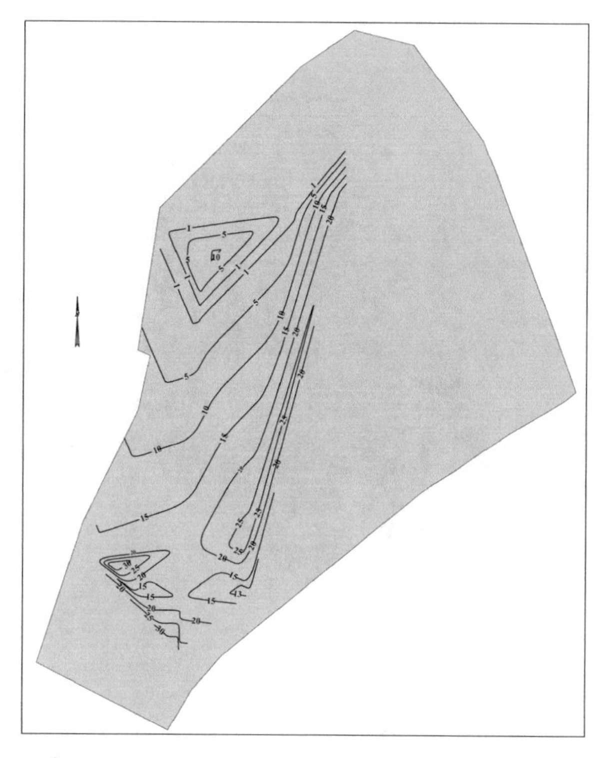

Fig. 5. Contour map of LPI for SPT results using reliability based probabilistic approach (Idriss and Boulanger model)

conservative as they gave more area with high liquefaction susceptibility (LPI > 15) than the deterministic methods. However, computing LPI based on the safety factor calculated according to the deterministic method of Youd et al. (2001) is the least conservative since it doesn't lead to zones with LPI higher than 15.

Figures 6 and 7 show the results of LPI contour maps using CPT results in conjunction with deterministic and reliability based probabilistic approach. A slight difference is observed between the LPI values based on deterministic and probabilistic approaches. Nevertheless, this slight difference shows that using the probability of liquefaction gives higher values of LPI than using the safety factor.

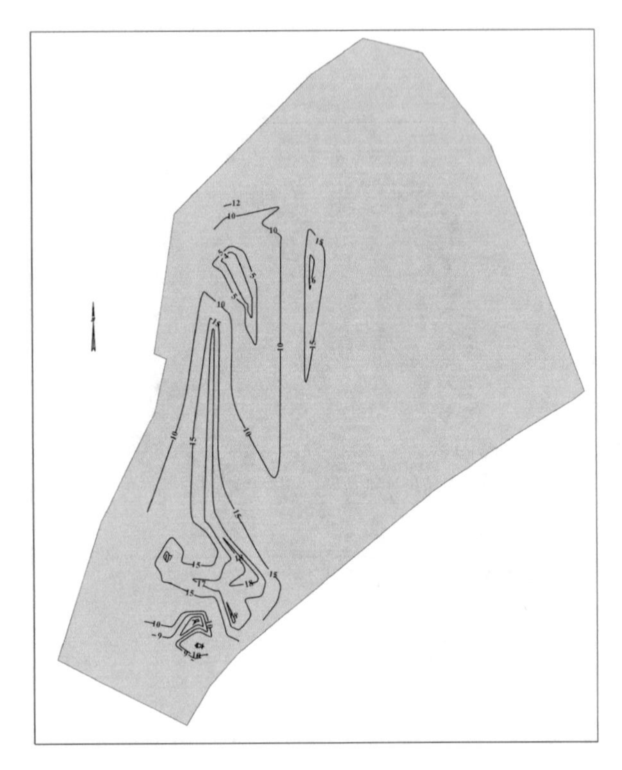

Fig. 6. Contour map of LPI for CPT results using deterministic approach (Idriss and Boulanger model)

Besides, both SPT and CPT results indicate that the southern part of the Lot is highly vulnerable to severe liquefaction with LPI values higher than 15. However, based on the SPT results, liquefaction is unlikely to occur in the northern part of the site with LPI lower than 5. On the other hand, by using the CPT results, the same zone has a moderate risk of liquefaction occurrence.

Given that the water table is approximately at the same level overall the site and that the same seismic parameters are considered, the difference in the LPI values can be directly attributed to the thickness of the potentially liquefiable soil and the relative density of the soil.

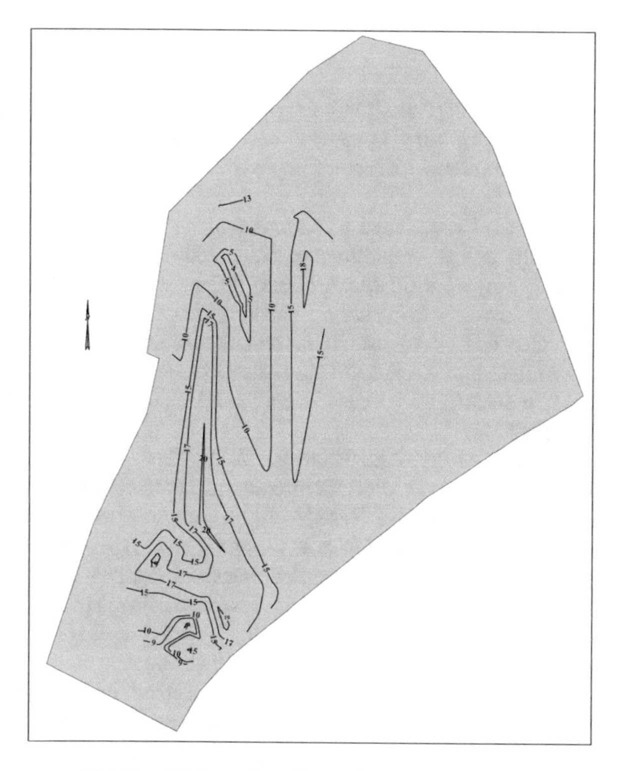

Fig. 7. Contour map of LPI for CPT results using reliability based probabilistic approach (Idriss and Boulanger model)

6 Conclusions

The evaluation of soil liquefaction is given by terms of safety factor or probability of liquefaction for a single soil element. Liquefaction potential index (LPI) is used to express the liquefaction severity for a whole soil profile.

Using SPT and CPT results and with conjunction with deterministic and reliability based probabilistic approaches, the calculation of LPI was conducted for the forty-nine executed boreholes in a studied area in the northern coast of Beirut. The spatial distribution of soil liquefaction potential is presented by means of contour maps for each considered method of calculation and for both in situ tests. The results show that the evaluation of LPI using the reliability based probabilistic approach and the CPT results is a more conservative method giving larger area with LPI values higher than 15.

In this paper, mapping was limited to one site on the coastal area where a potentially liquefiable soil exists. A continuity of this work will be to extend the analysis for Beirut City resulting in LPI contour mapping for the whole city based on probability of liquefaction using a reliability approach and the available SPT and/or CPT data.

References

Boulanger, R.W., Idriss, I.M.: CPT and SPT- Based liquefaction triggering procedures. Report No. UCD/CGM-14/01, April 2014. Center for Geotechnical Modeling– Department of Civil and Environmental Engineering– College of Engineering – University of California at Davis, USA (2014)

Cetin, K.O.: Reliability-based assessment of seismic soil liquefaction initiation hazard, PhD. Dissertation. University of California, Berkeley, CA (2000)

Cetin, K.O., Seed, R.B., Kayen, R.E., Moss, R.E.S., Bilge, H.T., Ilgac, M., Chowdhury, K.: Summary of SPT based field case history data of Cetin database. Report No. METU/CTENG08/16-01, August 2016. Soil Mechanics & Foundation Engineering Research Center- Middle East Technical University- Ankara, Turkey (2016)

DRM, World Institute for Disaster Risk Management (2003). Microzonation for Earthquake Risk Mitigation (MERM) (2003)

Holzer, T.L., Bennett, M.J., Noce, T.E., Padovani, A.C., Tinsley, J.C.: Liquefaction hazard mapping with LPI in the Greater Oakland, California, area. Earthq. Spectra **22**, 693–708 (2006)

Iwasaki, T., Tokida, K., Tatsuko, F., Yasuda, S.: A practical method for assessing soil liquefaction potential based on case studies at various sites in Japan. In: Proceedings of 2nd International Conference on Microzonation, San Francisco, pp. 885–896 (1978)

Iwasaki, T., Tokida, K., Tatsuoka, F., Watanabe, S., Yasuda, S., Sato, H.: Microzonation for soil liquefaction potential using simplified methods. In; 1982 Proceedings of 2nd International Conference on Microzonation, Seattle, pp. 1319–1330 (1982)

Juang, C.H., Fang, S.Y., Khor, E.H.: First-order reliability method for probabilistic liquefaction triggering analysis using CPT. J. Geotech. Geoenviron. Eng. ASCE **132**(3), 337–350 (2006)

Juang, C.H., Fang, S.Y., Tng, W.H., Khor, E.H., Kung, G.T.C., Zhang, J.: Evaluating uncertainty of an SPT-based simplified method for reliability analysis for probability of liquefaction Soils and Foundations, vol. 46, no 1, pp. 135–152, February 2009. Japanese Geotechnical Society (2009)

Luna, R., Frost, J.D.: Spatial liquefaction analysis system. J. Comput. Civil Eng. **12**, 48–56 (1998)

Moss, R.E.S.: CPT-based Probabilistic Assessment of Seismic Soil Liquefaction Initiation, Ph.D. dissertation. University of California, Berkeley, California (2003)

Sebaaly, G., Rahhal, M.E.: Model uncertainty of SPT based methods for reliability analysis of soil liquefaction. In: 7th International Conference on Earthquake Engineering, pp. 4906–4913 (2019)

Sebaaly, G., Rahhal, M.E.: Probabilistic analysis of soil liquefaction based on CPT and SPT Results. In: 7th ECCOMAS Thematic Conference on Computational Methods in Structural Dynamics and Earthquake Engineering (2019, in press)

Seed, H.B., Idriss, I.M.: Simplified procedure for evaluating soil liquefaction potential. J. Soil Mech. Found. Div. ASCE **97**(9), 1249–1273 (1971)

Sonmez, H.: Modification of the liquefaction potential index and liquefaction susceptibility mapping for a liquefaction-prone area (Inegol, Turkey). Environ. Geol. **44**, 862–871 (2003)

Toprak, S., Holzer, T.L.: Liquefaction potential index: Field assessment. J. Geotech. Geoenviron. Eng. **129**, 315–322 (2003)

Youd, T.L., Idriss, I.M., Andrus, R.D., Arango, I., Castro, G., Christian, J.T., Dobry, R., Liam Finn, W.D.L., Harder Jr., L.F., Hynes, M.E., Ishihara, K., Koester, J.P., Liao, S.S.C., Marcuson III, W.F., Martin, G.R., Mitchell, J.K., Moriwaki, Y., Power, M.S., Robertson, P. K., Seed, R.B., Stokoe II, K.H.: Liquefaction resistance of soils: summary report from the 1996 NCEER and 1998 NCEER/NSF workshops on evaluation of liquefaction resistance of soils. J. Geotech. Geoenviron. Eng ASCE **127**(10), 817–833 (2001)

Impact Analysis of Soil and Water Conservation Structures- Jalyukt Shivar Abhiyan- A Case Study

Ajay Kolekar[1], Anand B. Tapase[2(✉)], Y. M. Ghugal[3], and B. A. Konnur[1]

[1] Department of Civil Engineering, Government College of Engineering, Karad, Karad, Maharashtra, India
ajaykolekar1311@gmail.com, bakonnur@gmail.com
[2] Department of Civil Engineering, Rayat Shikshan Sanstha's, Karmaveer Bhaurao Patil College of Engineering, Satara, Maharashtra, India
tapaseanand@gmail.com
[3] Department of Applied Mechanics, Government College of Engineering, Karad, Karad, Maharashtra, India
yuwraj.ghugal@gcekarad.ac.in

Abstract. In the state of Maharashtra, civilians from 188 talukas were facing the drought-like situations till 2014–15. The groundwater level was lowered by 2–3 m due to inadequate and uncertain rainfall. To overcome the situation, the state government started to water and soil conservation works under Jalyukt Shivar Abhiyan scheme. In rural areas, various works like chain cement Nala bandh, desilting the reservoirs, repairs to K. T. weir bandhara, deep continuous contour trenches, compartmental bunding were carried out. As per the information received from the Water Conservation Department, more than 8000 crores spent in the last 4 years on this project. The expenditure is made village wise wherein from the data obtained it is noted that in around 15460 villages have received fund and 100% works were found complete, 80% works in 821 villages were found complete, 50% of works were found complete in 410 villages, 30% of works were found complete in 2,922 villages but still 2395 have still to start the work. The paper focuses on correlating the funds spend and the impact of various soil and water conservation works. Sample villages were assessed by conducting a survey at ground level. It is observed that in the number of villages the groundwater table was raised resulting in charging the wells from the nearby area. Conventional crop pattern was found improved with increased yield.

1 Introduction

The Jalyukt Shivar Abhiyan (JSA) is a program of Govt. of Maharashtra launched in 2015, as per the Government Resolution (G.R.) dated 5th December, 2014. The program aims to tackle the question of recurring droughts and make Maharashtra drought-free by the year 2019. The program is being implemented at village level and is supposed to cover all the villages in five years in a phased manner. Every year few

H. El-Naggar et al. (Eds.): GeoMEast 2019, SUCI, pp. 47–53, 2020.
https://doi.org/10.1007/978-3-030-34252-4_5

villages in each taluka are selected for the implementation based on pre-defined criteria (i.e. existing incomplete projects, drinking water scarcity, low agricultural productivity, groundwater exploitation etc.). Around 5000 villages are selected each year.

JSA is watershed program with main focus on soil and water conservation activities like trenches, gabions, percolation tanks, cement bunds, nala-deepening, farm level soil conservation activities and so on. Through these activities the program strives to conserve and harvest as much rain-water within village boundary as possible and resolve the problems of water stress during dry spells in monsoon season, shortage of water during rabbi season and drinking water scarcity, especially during summer season.

The planning and implementation of the above works is done through convergence of funds from all existing state and centre-level watershed programs as well as from MP, MLA and CSR funds. The planning at the village level has to be carried out in co-ordinated manner i.e. with the help of all concerned departments (Agriculture, Forest, Minor Irrigation, GSDA, RWS and so on) and the plan is to be discussed in the Gram Sabha. The District Collector is supposed to oversee the implementation of the program at the district-level.

One of the requirements of the program is that all the villages should be assessed after the completion of works in the village. This assessment is to be done by third-party and is supposed to be overseen by the District Collector. Currently many organizations are doing the assessment work. However, assessment reports of different agencies (some of them being local NGOs, regional academic institutes etc.) are of varying quality and content. Hence, it has been felt that a common guideline should be evolved and used for the assessment work across the state. This document should serve as such a guideline for the village level assessment of JSA. The document consists of methodology for assessment, survey formats, planning and steps to be followed, post-visit analysis and finally report writing.

2 Aims and Objectives of JYS Campaign

Considering drought-like situation occurring frequently in the state, Jalyukt Shivar campaign is being taken up under 'water for all - drought-free Maharashtra 2019':-

1. Harvesting maximum rainwater in the surrounding of village itself.
2. Increasing level of groundwater.
3. Increasing area under irrigation in the state - Increasing assured water for farming and efficiency of water usage.
4. Guaranteeing availability of sufficient water for all in the state - Increasing water supply by resurrecting dead water supply schemes in the rural area.
5. Implementing groundwater act. Creating decentralized water storages.
6. Initiating new projects to create water storage capacity.
7. Reinstating/increasing water storage capacity of existing and dysfunctional water sources (small dams/village tanks/percolation tanks/cement dams).
8. Extracting sludge from existing water sources through public participation and increasing water storage of water sources.

9. Encouraging tree plantation and planting trees. Creating public concern/awareness about balanced use of water.
10. Encouraging/creating awareness about efficient utilization of water for farming.
11. Sensitizing people about water harvesting/increasing public participation.

3 Background

3.1 State Profile

Maharashtra's total geographical area is 307.70 lakh hectares of which 225.4 lakh hectares area is "cultivable land". Maharashtra occupies the western and central part of the country and has a long coastline stretching nearly 720 km along the Arabian Sea. The Sahyadri mountain ranges provide a physical backbone to the State on the west, while the Satpuda hills along the north and Bhamragad-Chiroli-Gaikhuri ranges on the east serve as its natural borders. The State is surrounded by Gujarat to the North West, Madhya Pradesh to the north, Chhattisgarh to the east, Andhra Pradesh to the south east, Karnataka to the south and Goa to the south west. The state has a geographical area of 3, 07,713 km^2 and is bounded by North latitude 15°40′ and 22°00′ and East Longitudes 72°30′ and 80°30′. About 75% area of Maharashtra is drained by eastward flowing rivers; viz. the Godavari and Krishna, to the Bay of Bengal and the remaining 25% area is drained by westward flowing rivers like the Narmada, Tapi and Konkan coastal rivers to the Arabian Sea. Maharashtra is prone to various disasters such as drought, floods, cyclones, earthquake and accidents. While low rainfall areas of the state are under the constant risk of droughts, high rainfall zones of eastern and western Maharashtra are prone to flash floods and landslides. The Koyna reservoir and surroundings fall under the high risk of earthquake hazards. The Government of Maharashtra has established a mechanism for disaster preparedness and mitigation by integrating science and technology with communication network facilitates. The Deccan plateau constitutes 50% of the drought-prone areas of the state. 12% of the population lives in drought-prone areas. Once in 5 years, deficient rainfall is reported. Severe drought conditions occur once every 8–9 years. The 1996 drought affected 7 districts and 266.75 lakh people. The 1997 drought affected 17 districts. About 50% of the drought prone areas of Maharashtra are in the Deccan Plateau. About 90% of the land in the state has basaltic rock, which is non-porous and prevents rainwater percolation into the ground and thus makes the area drought prone.

3.2 Local Government Structure

The State has 35 districts which are divided into six revenue divisions viz. Konkan, Pune, Nashik, Aurangabad, Amravati and Nagpur for administrative purposes. The State has a long tradition of having statutory bodies for planning at the district level. For local self-governance in rural areas, there are 33 Zilla Parishads (District Councils), 351 Panchayat Samitis (Block Councils) and 27,906 Gram Panchayats (Village Council). The urban areas are governed through 26 Municipal Corporations, 222 Municipal Councils, 7 Nagar Panchayats (Notified Area Council) and 7 Cantonment Boards.

3.3 The Need to Conserve Water

Factually, Maharashtra has been bestowed with adequate rainfall, perennial rivers, lakes and large streams. However, due to concretization in last few decades, natural resources in the State have endured huge losses, driving it to situations like drought. Any type of natural activity is in essence completely balanced. Natural streams are created as a cumulative result of various land strata such as hills and hillocks, slightly deeper stretches, plateaus and grounds as well as green cover and rainfall in the area. The water bodies, in the form of rivers, streams, nallas and smaller streams, decide the sustainability and future of the region. Urbanization in any part of the land requires changes in the natural landscape of the region. The speed of urbanization in the State too adversely affected the water bodies it had. Today, one cannot find a city, town or a village in Maharashtra where natural streams have not been encroached. One of the major responsibilities undertaken through Jalyukt Shivar Abhiyan is to rejuvenate these natural water sources.

4 Data and Methods

Since last two years, chain cement concrete canal construction program, and various other water and land conservation campaign have been implemented in the state. Similarly, water harvesting activities like sludge extraction in Latur district and well refilling in Nanded district have been successfully conducted. To permanently overcome drought situation, Jalyukt Gaav (water full village) campaign was implemented in 5 districts from Pune division in the year 2012–13. Under this, action plan was prepared for water harvesting and increasing groundwater level by implementing various schemes collectively through coordination of all departments. Activities taken up under this campaign were - watershed projects in the division through water conservation, cement chain canal constructions, repair and renovation of old existing cement canal dams/K.T. Weir, sludge extraction from water source, water source empowerment, well refilling, efficient utilization of available water and canal joining works. Through all these projects, decentralized water storage of 8.40 TMC capacities has been created. Because of this, groundwater level is increased by 1 to 3 m and provision for drinking water and protected irrigation for farming is made. This has helped to permanently overcome drought situation. Considering results of all these projects, the government was thinking of preparing organized action plan to make ‘water for all - drought-free Maharashtra and to permanently overcome drought situation and implementing ‘Jalyukt Shivar’ (water full surrounding) campaign to increase water availability.

After witnessing the tremendous works planned and carried out in drought-affected villages, people were encouraged to participate, and within seven months, a total of 99,154 works were completed in 6,202 villages. The scheme helped villagers realize what they had lost over the years. To create further awareness among people, water processions were organized in various villages. In the backdrop of a celebratory mood set by these processions, the villages keenly looked at the subject, and awareness was created. Small groups took the lead and achieved participation of other villagers.

A. Impetus to Jalyukt Shivar Campaign
   Under Jalyukt Shivar Abhiyan works of compartment bunding, de-siltation, mud nulla (lake) bunding, cement check dams, repairs of bunding, KT weirs, farm ponds and village talav, refilling of wells, recharge shaft, drip irrigation and sprinkle irrigation have been brought together.
A. What Lies Ahead?
   After completion of irrigation projects in next two years, 50% area will be under irrigation. For the rest 50%, rainwater harvesting and decentralizing water sources are the only options to solve the issue of water scarcity. The Chief Minister has made a suggestion to carry out a structural mapping of the work done so far. He has set a target of 13,000 villages in two years and 20,000 villages in five years to be draught-free. He believes that to achieve this, rainwater harvesting is the only way forward. As the first phase of the scheme is successfully near completion at most places, the CM now wants to emphasize on correct crop patterns, group farming and forming farmers' cooperatives in the next phase. From next year, 25% funds will be allocated for maintenance and repair work of these projects and all water sources will be brought under the scheme, including rivers.

5 Benefits Due to JYS Campaign on Practical Implementation

A. Rivers Deepened, Water Level In The Wells Enhanced The results of bending nature according to our whims are best explained by the villagers of Telhara, a small hamlet in the foothills of Satpuda range. With good precipitation, agriculture has been flourishing here. However, every year during the rains, the villagers used to be on their toes. River Gautama flows near the village and there was a huge encroachment on the banks. Trees were proliferating in the river bed and the river had become shallow. Every rainy season, Telhara used to get waterlogged and the farmers suffered huge losses. During the floods in 2014–15, an area of 383 ha around the village was affected. To overcome the problem, there was a need to deepen the river bed. People joined hands and a stretch of 10 km was deepened and widened. Sediment to the extent of 5.4 lakh cu.m was removed. The farmers made use of the sediment to make their soils fertile.
B. Water-Filled Dams, Healthy Crops In Karveer taluka of Kolhapur, the construction of a cement nalla bunding began in January this year. By July, the construction was done and the water stored in the dam was helping in the irrigation of the region's sugarcane crop. Karveer today presents an opportunity to witness the huge change water can bring about in the life of people, especially farmers. Even after the absence of rain in June and before it started intensely, the benefits of Jalyukt Shivar Abhiyan to the villages were very much visible.
C. Use of Mobile App and Representation on GIS Platform Proper description of the situation is one of the most important first steps for any program to succeed. This requires representation and proper analysis of all the relevant data. This demands for proper data collection of various parameters like terrain, soil, geology, rainfall,

wells, crops, quality and so on, in order to tackle the problem of water security. In order to have proper representation and analysis, this data needs to be transported to GIS platform. JYS GR (dated 5-12-2014) mentions the importance and use of GIS in planning and representation. Currently most of the data like soil, geology, land use etc. is available with Maharashtra Remote Sensing Application Centre (MRSAC) as GIS shape files at village level. The mobile app developed by MRSAC, is being used to these locations. The mapped location can be monitored through web page. These GIS layers can be very useful in understanding the nature and causes of drinking water scarcity, quality problems, impact of conservation structures etc. Use of GIS also makes available different maps like drinking water stress maps, quality affected areas maps, sugarcane belts, poor groundwater belts etc. Such maps would serve two purposes; (i) maps convey more information than tables and reports, hence villagers will become more aware and (ii) these maps give further direction in understanding the problem better.

6 Broader Suggestions

A. Interaction and Coordination between Departments- Proper representation of data on GIS platform requires integration of data from different departments. Different datasets like revenue and land use data from Revenue department, crop data from Agriculture department, canal and command area data from Water Resources department, groundwater assessment data from GSDA, conservation structures data from Soil conservation department, watershed data from IWMP (Agriculture department), drinking water data from Water Supply and Sanitation department etc. has to be brought to one place for correct analysis and formulation of the problem. This requires proper integration and communication between all these departments. JYS GR mentions this as a requirement while preparing all the village plans. But there is no clear provision and room to make such interaction and communication in the village planning framework. Village plans talk about financial convergence between various departments and programs, but this need to be extended to convergence of data, capacities and so on.

B. Groundwater Modelling and Simulations- Some complex problems might require more research and analysis and use of tools such as groundwater modeling for greater understanding of the problem. For example, finding suitable areas for interventions like lake-deepening would include understanding of the geology, aquifer characteristics and groundwater flows. Similarly, impact of recharge shafts or identification of source of contamination of drinking water in villages etc. can be carried out by using groundwater modeling and simulations. A suite of such simulations will help in designing JYS better.

7 Conclusions

Jalyukt Shivar Campaign has become a people's movement in Maharashtra, and it is proving to be useful for irrigation and enhancement of groundwater level. This program should be strictly implemented as a campaign through government departments, voluntary organizations, public participation and funds available with private businessmen (CSR), to ensure drought does not occur in future in drought prone areas and remaining area. In order to create public awareness it is important that along with the JYS promotion-vehicle travelling to villages, village maps showing all the proposed and existing interventions should be displayed in the Gram Panchayat (Village Council) office and schools. Implementing Jalyukt Shivar Campaign in other parts of country will be helpful to overcome permanently drought and water scarcity problems. With unique initiative like Jalyukt Shivar, water scarcity will surely be a thing of the past!

References

1. A report on Watershed Interventions for Kurlod and Botoshi Phase-II, Technology and Development Solutions Cell (TDSC) Centre for Technology Alternatives for Rural Areas (CTARA), Indian Institute of Technology, Bombay (IITB), July 2015
2. A report on Watershed Interventions for Kurlod and Botoshi Phase-I, Technology and Development Solutions Cell (TDSC) Centre for Technology Alternatives for Rural Areas (CTARA), Indian Institute of Technology, Bombay (IITB), December 2014
3. Hutti, B., Nijagunappa, R.: Applications of geo informatics in water resources management of semi-arid region, North Karnataka, India. Int. J. Geomatics Geosci. **2**(2), 374–382 (2011)
4. Saran, S., Mittra, S., Hasan, S.: Attitudes towards water in India. Observer Research Foundation, June 2014
5. Government of Maharashtra, Water Conservation Department, Government Resolution No. JaLaA-2014/Case No. 203/JaLa-7, Mantralaya, Mumbai - 400 032, 5 December 2014
6. Government of Maharashtra, Planning Department, Report of The High Level Committee on Balanced Regional Development Issues in Maharashtra, October 2013
7. Lalbiakmawia, F.: Application of geo-spatial technology for ground water quality mapping of Mamit district, Mizoram, India. IJESMR **2**, 31–38 (2015)

Failure of Overhead Line Equipment (OHLE) Structure Under Hurricane

Chayut Ngamkhanong[1], Sakdirat Kaewunruen[1(✉)], Rui Calçada[2], and Rodolfo Martin[3]

[1] School of Engineering, University of Birmingham, Birmingham, UK
s.kaewunruen@bham.ac.uk
[2] Faculty of Engineering, University de Porto, Porto, Portugal
[3] Evoleo Technology Pty Ltd., Porto, Portugal

Abstract. Presently, in modern railway systems, train or rolling stocks are powered by electricity through the overhead wire or the third rail on ground. Overhead line equipment (OHLE) is the component for the electric train which provides electric power to the train. OHLE is, for one or two tracks, normally supported by cantilever masts. OHLE is one of vulnerable components in railway system due to its slenderness. Note that, as previously recorded, the strong hurricane caused substantial damage over the large area and possibly knocked the train out of the track and cause electricity failures on OHLE. In fact, cantilever mast subjected to wind and hurricane actions may fail due to the incorrect design, material defects, improper support connections and its foundation etc. In this study, a mast structure with varying rotational soil stiffness is used to construct dynamic influential lines for soil-structure integrity prediction. Finite element model updating technique has been used to perform the dynamic responses of OHLE considering soil-structure interaction of OHLE. The scaled hurricanes at various magnitudes are applied to the OHLE. It is interesting that the support condition plays a significant role in the dynamic responses of OHLE under strong hurricanes. The obtained results demonstrate that the strong hurricane can cause a catastrophic damage to the OHLE which is linked to the failure of electric train. The insight will raise the awareness of engineers for better design of cantilever mast structure and its support condition.

1 Introduction

Overhead Line Equipment (OHLE), which consists of masts, gantries, wires found along electrified railways, is the component for the electric train to supply power to make electric trains move, It is important that electric train has become the efficient railway system that emits less carbon and is allowed to run quicker and more frequent for the sudden growth of passengers and journeys (RailCorp 2011). OHLE is supported by the mast structures located alongside the railway line. However, due to the loss of contact between pantograph and contact wire, the electric system may be failed. There are some reasons that can possibly make a loss of contact wire such as wind, temperature, broken tree, snow storm, train derailment, track buckling etc. (Shing and Wong 2008; Robinson and Bryan 2009; Taylor 2013; Beagles et al. 2016).

H. El-Naggar et al. (Eds.): GeoMEast 2019, SUCI, pp. 54–63, 2020.
https://doi.org/10.1007/978-3-030-34252-4_6

Due to the effects of global climate change, all the natural disasters tend to have higher intensity and occur more frequent. It should be noted that strong wind and hurricane can knock the structure down as can be seen in many evidences occurred such as billboard, lattice tower, lighting pole etc. (Letchford 2001; Tamura and Cao 2009; Ramalingam 2017; Li et al. 2018). As for railway track infrastructure, if the wind becomes strong enough to form a hurricane, the trees next to the train lines might fall and hit the overhead contact wires or knock down the mast structure (Network Rail 2017), thereby leading to the failure of the electrical power system. For one or two railway tracks, OHLE is normally supported by single cantilever mast structure. It is interesting to note that the main reasons of damage and failure of mast structure are the poor of support structure, such as broken bolt, yielding weld, improper design and construction, and the dynamic sensitivity due to its slenderness by nature leading to the loss of contact wire.

The cantilever mast is a vibration-sensitive structure since crossing phenomena can be observed when the support stiffness is changed (Ngamkhanong et al. 2017a, b). It was confirmed by previous studies that soil-structure interaction affected the overall response of the structure (Prum and Jiravacharadet 2012; NEHRP 2012).

Due to the change of global climate, hurricane or strong wind may cause damage to the OHLE structure by losing the contact between pantograph and contact wire which can lead to operational failure of train electrification. The integrated numerical study of three-dimensional cantilever mast structure under the strong wind is presented to evaluate the condition of OHLE structures for maintenance planning. The condition of OHLE can be monitored using adaptation technique by finite element package STRAND7 (G+D Computing 2001). Wind actions are calculated by the formulation based on the standard of wind actions on structures. The lateral displacement of the contact wire on the cantilever is measured to be able to properly inspect the contact wire condition. The soil-structure interaction is also taken into account. This study presents the maintenance index which can be used for maintenance planning and inspection of support condition of mast and OHLE structures. The outcome of this study will help civil and track engineers to effectively and efficiently inspect OHLE structures and its support using the structural response from wind actions.

2 Methodology

2.1 Modelling

Three-dimensional modelling of single mast structure is constructed using finite element package STRAND7 (G+D Computing 2001). The linear static solver is used based on the assumprion that structure is in linear range and loading is static. The single cantilever mast structure made by steel is constructed, Fig. 1. The 2-D schematic load to structure with support springs is shown in Fig. 2. The typical H-section steel (Section area: 2.219×10^{-2} m^2, I_{zz}: 5.08×10^{-4} m^4, I_{xx}: 1.84×10^{-4} m^4) is used and connected to the cantilever which made of round steel to support the overhead contact wire. The steel used has the young modulus of 2×10^5 MPa, density of 7850 kg/m^3 and poisson's ratio of 0.25. The rotational and translational springs are applied at the support. The parametric study of the soil stiffness is conducted. It is assumed that translational stiffness of support is fully fixed in all directions, while the rotational stiffness is varied from 500 to 1000000 kNm/rad (fully fixed support condition). It is noted that the rotational stiffness is affected by the soil-structure interaction condition and the quality of support connection (Kanvinde et al. 2012; Rodas et al. 2017; Krystosik 2018). The support stiffness can be decrease due to the connection failure such as broken bolt, yielding weld, improper design and construction etc. and soil erosion and degradation. It is noticeable that the soil-structure interaction plays a significant role in the sensitivity of structural vibration. The quality of support connection and ground condition considerably influences the soil-structure stiffness (Ngamkhanong et al. 2017a, b) (Fig. 3).

Fig. 1. Single mast structure supporting OHLE

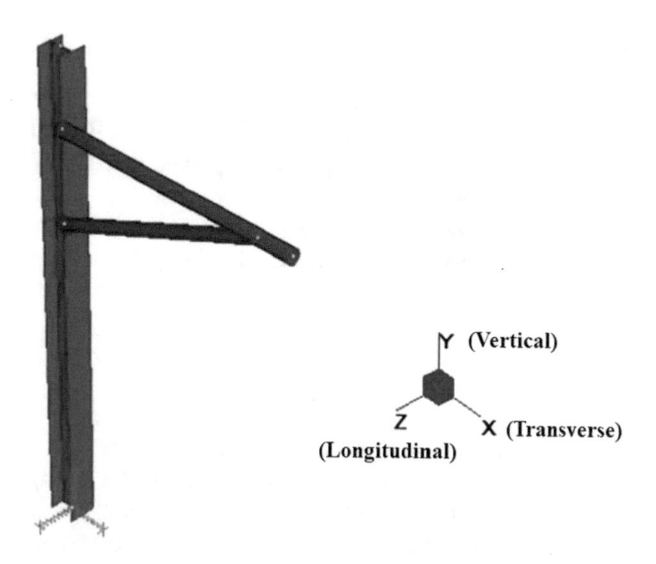

Fig. 2. 3-Dimensional model of OHLE

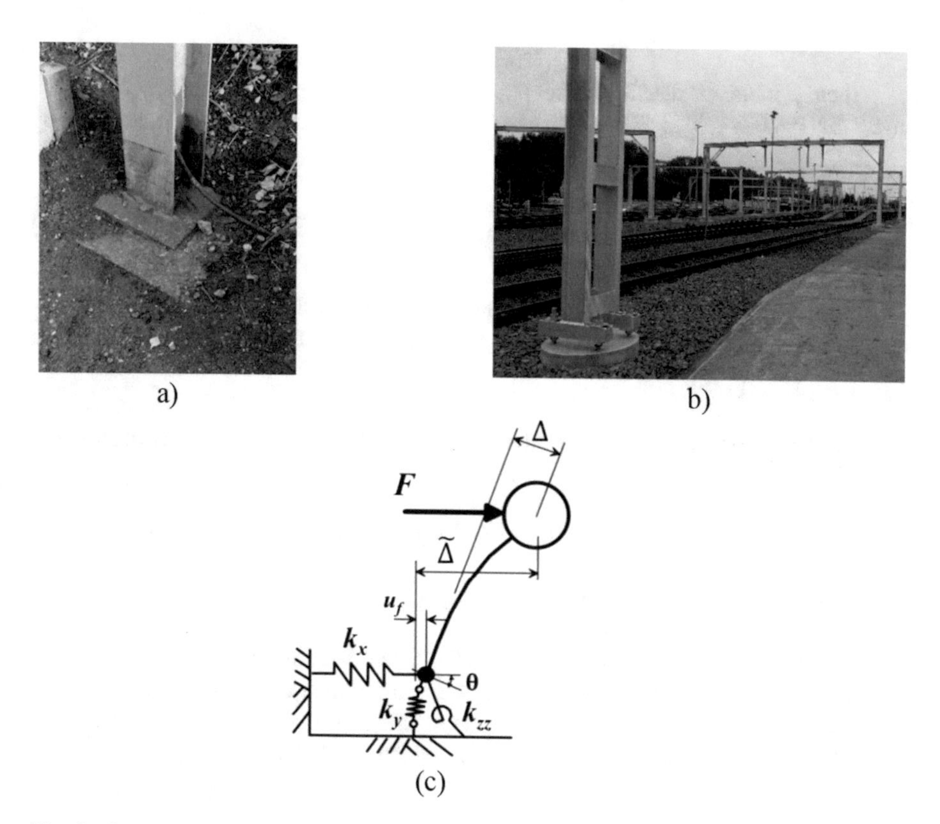

Fig. 3. Support of (a) cantilever mast (b) frame mast and (c) Schematic load to structure with rotational flexibility at support

2.2 Wind Action Calculation

The wind action calculated from the wind velocity is presented by a static pressures or forces acting on the face of structure (BSI 2005; BSI 2006a, b; BSI 2012). The wind actions are characteristic values depending on the type and location of structure. The steps of wind force calculation are shown below.

The turbulence intensity and mean wind pressure at the reference height above ground, h, can be calculated using Eqs. 1 and 2, in order to further compute the peak wind pressure.

$$I_V(h) = 1/[c_0 \ln\left(\frac{h}{z_0}\right)] \tag{1}$$

Where c_0 is the orography factor, z_0 is the roughness length.

Note that the recommended value is 1 for both c_0(when the average slope of the upwind terrain is small) and z_0(value for the area in which at least 15% of the surface is covered with buildings which average height exceeds 15 m) as the worst case scenario.

$$q_h(h) = \frac{1}{2}\rho V_h^2(h) \tag{2}$$

Where ρ is the air density (kg/m^3). A conservative value for ρ is 1.25 kg/m^3 given in BS EN1991-1-4 (BSI 2005). $V_h(h)$ is the mean wind velocity (m/s) calculated by Eq. 3.

$$V_h(h) = V_{b,0} c_{dir} c_o k_r \ln(\frac{h}{z_0}) \tag{3}$$

Where $V_{b,0}$ is basic wind velocity (m/s), c_{dir} is wind directional factor (The recommended vale is 1), c_o is orography factor (taken as 1), k_r is the terrain factor $\left(k_r = 0.19\left(\frac{z_0}{z_{0,II}}\right)^{0.07}\right)$.

The peak wind pressure at the reference height above the ground h is determined using Eq. 4.

$$q_p(h) = [1 + 7I_V(h)]q_h(h) \tag{4}$$

Where $I_V(h)$ is the turbulence intensity.

$q_h(h)$ is the mean wind pressure.

Based on BS EN 50341-1:2012 (BSI 2012), wind action on the overhead line is basically calculated by the wind pressure acting multiplied by the projected area and structural factors. In this case, the wind force acting on mast structure can be determined by the Eq. 5 based on the assumption of the wind forces acting on poles.

$$Q_w = q_p(h)GCA \tag{5}$$

Where G is the structural factor, the recommended value is 1. C is the drag factor (1.8 is used for steel with sharp edge cross section). A is the projected area on a vertical plan perpendicular to the wind direction.

2.3 Hurricane Wind Speed

The scale was developed in 1971 by civil engineer Herbert Saffir and meteorologist Robert Simpson, who at the time was director of the U.S. National Hurricane Center. The U.S. National Weather Service, Central Pacific Hurricane Center and the Joint Typhoon Warning Center define sustained winds as average winds over a period of one minute, measured at the same 33 ft (10.1 m) height and that is the definition used for this scale (Federal Emergency Management Agency 2004; Tropical Cyclone Weather Services Program 2006). These velocities are then used to calculate the static force to apply to the cantilever mast. Hurricane's sustained wind speed can be classified into 5 categories based on potential property damage, as shown in Table 1.

Table 1. Hurricane wind scale and types of damage (Tropical Cyclone Weather Services Program 2006)

Category	Wind speed	Types of Damage Due to Hurricane Winds
1	74–95 mph 64–82 kt 119–153 km/h	Very dangerous winds will produce some damage: Well-constructed frame homes could have damage to roof, shingles, vinyl siding and gutters. Large branches of trees will snap and shallowly rooted trees may be toppled. Extensive damage to power lines and poles likely will result in power outages that could last a few to several days
2	96–110 mph 83–95 kt 154–177 km/h	Extremely dangerous winds will cause extensive damage: Well-constructed frame homes could sustain major roof and siding damage. Many shallowly rooted trees will be snapped or uprooted and block numerous roads. Near-total power loss is expected with outages that could last from several days to weeks
3	111–129 mph 96–112 kt 178–208 km/h	Devastating damage will occur: Well-built framed homes may incur major damage or removal of roof decking and gable ends. Many trees will be snapped or uprooted, blocking numerous roads. Electricity and water will be unavailable for several days to weeks after the storm passes
4	130–156 mph 113–136 kt 209–251 km/h	Catastrophic damage will occur: Well-built framed homes can sustain severe damage with loss of most of the roof structure and/or some exterior walls. Most trees will be snapped or uprooted and power poles downed. Fallen trees and power poles will isolate residential areas. Power outages will last weeks to possibly months. Most of the area will be uninhabitable for weeks or months
5	157 mph or higher 137 kt or higher 252 km/h or higher	Catastrophic damage will occur: A high percentage of framed homes will be destroyed, with total roof failure and wall collapse. Fallen trees and power poles will isolate residential areas. Power outages will last for weeks to possibly months. Most of the area will be uninhabitable for weeks or months

Note: National Hurricane Center classifies hurricanes of Category 3 and above as major hurricanes, and the Joint Typhoon Warning Center classifies typhoons of 150 mph or greater (strong Category 4 and Category 5) as super typhoons.

The linear static analysis is used for wind action. The loads obtained by the calculation are applied to the mast structure in perpendicular direction. The nodal displacement at the cantilever, which is the location of contact wire, are taken into consideration and will be compared with the maintenance index. The ratio between the overhead contact wire displacement and allowable displacement is indicated as maintenance index. It is assumed and noted that the allowable displacement used is the construction tolerances of contact stagger above the track to avoid wearing a groove in the pantograph according to RailCorp (2011). The construction tolerance of 50 mm of contact wire is used as the allowable lateral displacement. Thus the maintenance index is the value between 0 and 1. The pantograph can possibly lost contact with the contact wire when the maintenance index reaches 1.

2.4 Results

The applied forces equivalent to the wind speed up to 300 km/h are applied to the mast structure. The displacement response of the mast structure in transverse direction subjected to the hurricane is presented in Fig. 4. The displacement response at the cantilever mast is then used to calculate the maintenance limit. The rotational stiffness considered is between 500 and 100000 kNm/rad (fully fixed support).

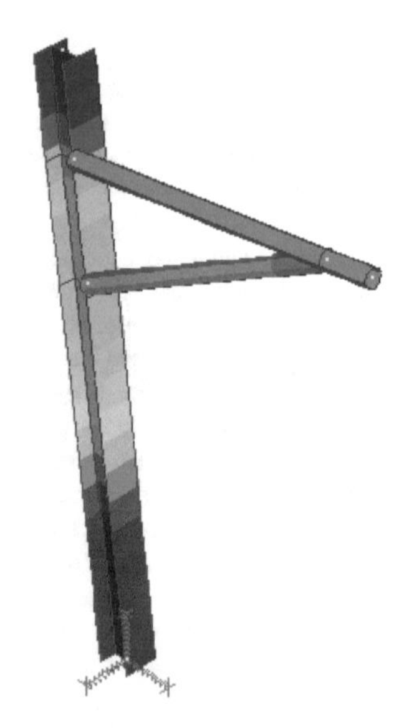

Fig. 4. Structural response of the mast structure subjected to hurricane.

The maintenance index of the OHLE at various stiffness conditions under different wind speed is presented in Fig. 5. It is clear that the strong wind or hurricane can significantly knock the electric system of the train. The results shown that the maintenance level can be reached when the rotational stiffness of the support is lower than 3000 kNm/rad which is represented by the poor support interaction such as low number of bolts, size of foundation, soil condition etc. It is noted that, the hurricane category one can possibly make a failure of the OHLE when the mast has the support stiffness of 500 kN/rad. However, the effect of strong wind can be mitigated by improving the support stiffness to be higher than 3000 kNm/rad. It is interesting to note that the strongest wind that has occurred, cannot fail the OHLE since the structure has a fully fixed support. The maintenance is only about 0.1 in this case.

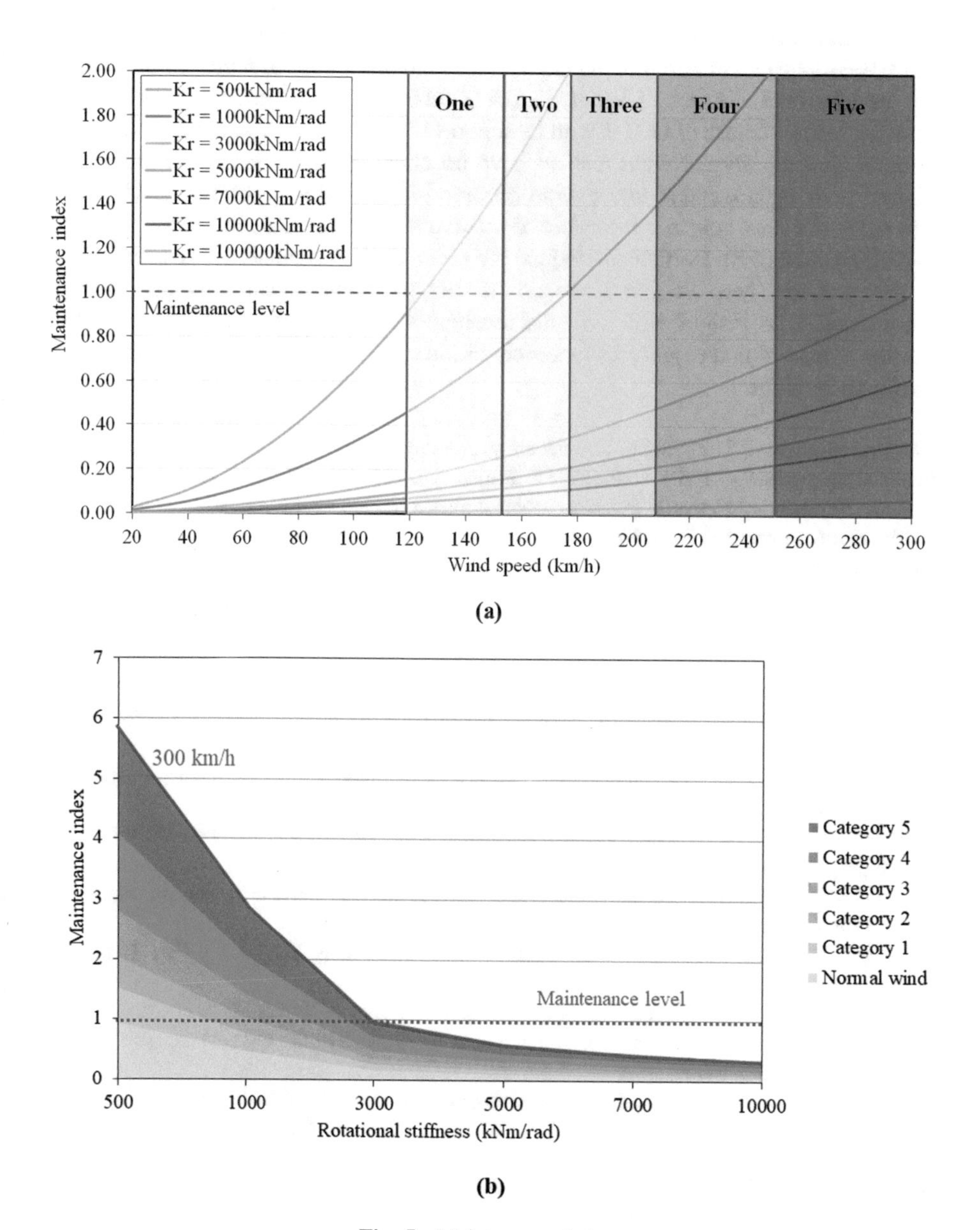

Fig. 5. Maintenance index

3 Conclusions

This study presents the structural responses of mast structure and OHLE under the hurricane and strong wind. The contact wire displacement on cantilever mast is measure based on the assumption that the structure can be failed when the displacement reaches the certain limit resulting in losing the contact between pantograph and contact wire. Moreover, the structure-soil interaction is considered as this can effectively affect the dynamic properties of the cantilever mast. The results obtained show that the failure of OHLE subjected to strong wind can be observed when the structure-soil interaction has its rotational stiffness of lower than 3000 kNm/rad. Hurricane category A could possibly lead to failure of OHLE with its support rotational stiffness of 500 kNm/rad. It is noted that the normal wind tend to have no effect on the failure of OHLE as the highest maintenance index under wind observed is around 90% of failure. However, this study does not consider the effect of relative displacement as, in reality, there is a gap at the contact area which can induce the relative displacement. This aspect will be further studied. Therefore, the connection at support should be carefully designed and constructed. The insight will raise the awareness of engineers for better design of cantilever mast structure and its support condition to encounter the natural disasters and future uncertainties.

Acknowledgments. The authors are sincerely grateful to the European Commission for the financial sponsorship of the H2020-RISE Project No. 691135 “RISEN: Rail Infrastructure Systems Engineering Network”, which enables a global research network that tackles the grand challenge of railway infrastructure resilience and advanced sensing in extreme environments (www.risen2rail.eu) (Kaewunruen et al. 2016).

References

Beagles, A., Fletcher, D., Peffers, M., Mak, P., Lowe, C.: Validation of a new model for railway overhead line dynamics. Proc. Inst. Civ. Eng. **169**, 339–349 (2016)

BSI: BS EN 1991-1-4:2005 Eurocode 1: Actions on structures — Part 1-4: General actions — Wind actions (2005)

BSI: BS EN 1993-3-1Eurocode 3: Design of steel structures - Part 3-1: Towers, masts and chimneys – Towers and masts (2006a)

BSI: BS EN 1993-3-2 Eurocode 3: Design of steel structures - Part 3-2: Towers, masts and chimneys – Chimneys (2006b)

BSI: BS EN 50341-1:2012 Overhead electrical lines exceeding AC 1 kV - Part 1: General requirements - Common specifications (2012)

Federal Emergency Management Agency: Hurricane Glossary of Terms (2004)

G + D Computing: Using Strand7: Introduction to the Strand7 finite element analysis system, Sydney, Australia (2001)

Haiderali, A.E.: Mitigation of ancient coal mining hazards to overhead line equipment structures. Proceedings of the Institution of Civil Engineers – Transport (2019). https://doi.org/10.1680/jtran.18.00143

Kaewunruen, S., Sussman, J.M., Matsumoto, A.: Grand challenges in transportation and transit systems. Front. Buil. Environ. **2**(4) (2016)

Kanvinde, A.M., Grilli, D.A., Zareian, F.: Rotational stiffness of exposed column base connections: experiments and analytical models. J. Struct. Eng. **138**(5), 549–560 (2012)

Krystosik, P.: Influence of supporting joints flexibility on statics and stability of steel frames. Int. J. Steel Struct. **18**, 433–442 (2018)

Letchford, C.W.: Wind loads on rectangular signboards and hoardings. J. Wind Eng. Ind. Aerod. **89**, 135–151 (2001)

Li, Z., Wang, D., Chen, X., Liang, S., Li, J.: Wind load effect of single-column-supported two-plate billboard structures. J. Wind Eng. Ind. Aerodyn. **179**, 70–79 (2018)

NEHRP Consultants Joint Venture: Soil-Structure Interaction for Building Structures (National Institute of Standards and Technology) (2012)

Network Rail: How stroms and flooding affect the railway (2017). https://www.networkrail.co.uk/storms-affect-railway-team-orangeprepares/

Ngamkhanong, C., Kaewunruen, S., Baniotopoulos, C., Papaelias, M.: Crossing phenomena in overhead line equipment (OHLE) structure in 3D space considering soil-structure interaction. In: IOP Conf. Series: Materials Science and Engineering, p. 245 (2017a)

Ngamkhanong, C., Kaewunruen, S., Baniotopoulos, C.: A review on modelling and monitoring of railway ballast. Struct. Monit. Maintenance **4**(3), 195–220 (2017b). https://doi.org/10.12989/smm.2017.4.3.195

Ngamkhanong, C., Kaewunruen, S., Baniotopoulos, C.: Far-Field earthquake responses of overhead line equipment (ohle) structure considering soil-structure interaction. Front. Built Environ. **4**(35) (2018a). https://doi.org/10.3389/fbuil.2018.00035

Ngamkhanong, C., Kaewunruen, S.: The effect of ground borne vibrations from high speed train on overhead line equipment (OHLE) structure considering soil-structure interaction. Sci. Total Environ. **627**, 934–941 (2018). https://doi.org/10.1016/j.scitotenv.2018.01.298

Ngamkhanong, C., Kaewunruen, S., Costa, B.J.A.: State-of-the-art review of railway track resilience monitoring. Infrastructures **3**, 3 (2018b)

Ngamkhanong, C., Kaewunruen, S., Calçada, R., Martin, R.: Condition monitoring of Overhead Line Equipment (OHLE) structures using ground-bourne vibrations from train passages. In: Rodrigues, H., Elnashai, A. (eds.) Advances and Challenges in Structural Engineering. GeoMEast 2018. Sustainable Civil Infrastructures. Springer, Cham (2019). https://doi.org/10.1007/978-3-030-01932-7_2

Prum, S., Jiravacharadet, M.: Effects of soil structure ınteraction on seismic response of buildings. In: International Conference on Advances in Civil Engineering for Sustainable Development (2012)

RailCorp: Design of Overhead Wiring Structures & Signal Gantries. Engineering Manual –Civil (2011)

Ramalingam, R.: Failure analysis of lattice tower like structures. In: IOP Conference Series Earth and Environmental Science, vol. 80, no. 1, p. 012024 (2017)

Robinson, P., Bryan, C.: Network rail electrical power reliability study. Network Rail, Milton Keynes (2009)

Rodas, P.B., Zareian, F., Kanvinde, A.: Rotational stiffness of deeply embedded column-base connections. J. Struct. Eng. **143**(8), 04017064 (2017)

Shing, A.W.C., Wong, P.P.L.: Wear of pantograph collector strips. Proc. ImechE, J. Rail Rapid Transit. **222**(2), 169–176 (2008)

Tamura, Y., Cao, S.: Climate change and wind-related disaster risk reduction. In: Proceedings of the APCWE-VII, Taipei, Taiwan 2009

Taylor, G.: A bad wire day. The Rail Engineer (2013)

Tropical Cyclone Weather Services Program: Tropical cyclone definitions (PDF). National Weather Service (2006)

Use and Comparison of New QA/QC Technologies in a Test Shaft

Patrick J. Hannigan(✉) and Rozbeh B. Moghaddam

GRL Engineers Inc., Cleveland, OH, USA
{phannigan, rmoghaddam}@grlengineers.com

Abstract. Drilled shafts are increasingly being used for foundation support. The quality of the constructed foundation is critical due the heavy foundation loads and limited redundancy of many drilled shaft foundations. On a recent project in the United States, several traditional and newer methods of quality control and quality assurance were used to assess the drilled shaft excavation, base condition, concrete quality, and capacity. The radii, shape, verticality, and volume of the drilled shaft excavation was evaluated with a SHaft Area Profile Evaluator (SHAPE), and the cleanliness of the shaft base prior to concrete placement was assessed with a Shaft Quantitative Inspection Device (SQUID). The placed concrete quality was evaluated with Cross-hole Sonic Logging (CSL) as well as Thermal Integrity Profiling (TIP). Finally, a bi-directional static load test (BDSLT) was conducted on the test shaft to determine the shaft capacity.

This paper will provide a brief review of the QA/QC tests and their results. In addition, the constructed shaft quality information available from similar methods will be compared and discussed including advantages and disadvantages of the respective methods.

1 Introduction

On drilled shaft projects in the U.S., a trial shaft and/or a test shaft is often specified to check the proposed shaft installation methods before proceeding into production. These trial or test shafts typically require quality assurance tests for base cleanliness, verticality, and concrete quality during construction. Trial shafts are used to evaluate the contractor's means and methods and are not statically loaded tested. A test shaft typically includes a conventional top-down or, more frequently, a bi-directional jack assembly so that the load carrying capacity can be assessed in additional to the other quality assurance tests.

Base cleanliness requirements vary depending upon the design and expected load transfer mechanism, the bearing materials, and whether the shaft was completed using wet or dry construction methods. The U.S. Federal Highway Administration guide specification for drilled shafts, Brown et al. (2018), limits sediment and debris thickness for wet or dry shafts in rock to less than 13 mm (0.5 in.) over 50% of the base area. For shafts on soil, the sediment and debris thickness is limited to less than 75 mm (3 in.) for wet construction, and less than 37 mm (1.5 in.) for dry construction.

H. El-Naggar et al. (Eds.): GeoMEast 2019, SUCI, pp. 64–87, 2020.
https://doi.org/10.1007/978-3-030-34252-4_7

More restrictive base cleanliness criteria are frequently specified when the shaft is expected to carry a large portion of the applied load through base resistance. In these cases, the average thickness of sediment is limited to 13 mm over 50% of the base area with no portion of the base having more than 37 mm (1.5 in.) of debris. Base cleanliness criteria are enforced for confirmation of base resistance and settlement considerations as well as to minimize possibility of concrete contamination from debris.

Shaft verticality or plumbness is often specified to be within 1.5% of plumb in soil and within 2.0% of plumb in rock, AASHTO (2017), Brown et al. (2018). Plumbness is measured from the top of shaft or from the mudline, whichever is lower.

The as-constructed quality and integrity of drilled shaft concrete is typically assessed through concrete volume plots versus elevation, cross-hole sonic logging test results, or thermal integrity profiling. Concrete volume versus elevation plots are described in Brown et al. (2018), guidance on evaluating integrity test results is available in an industry authored document for cross-hole sonic logging results by the Deep Foundations Institute (2019), and for thermal integrity profiling in Piscsalko et al. (2016). Guidance from these documents will be referred to later in this document.

2 Test Shaft Details

At the test shaft location, the general subsurface conditions consist of very loose silty fine sand (SM) to very soft sandy clay (CH) to a depth of 16.1 m (53 ft). These materials were underlain by a marl layer consisting of very loose to medium dense silty sand (SM) to clayey sand (SC) to a depth of 20.3 m (66.7 ft). The marl layer was in turn underlain by a limestone formation comprised of loose to medium dense silty fine sand (SM) with occasional strongly cemented sand layers. The test shaft was terminated in this limestone formation.

The test shaft was constructed by first vibrating a 22.3 m (73 ft) steel casing to a depth of 18.6 m (60.9 ft) below grade. The casing was 2,490 mm O.D. × 16 mm wall (98 in × 5/8 in). A 2,390 mm O.D. (94 in) Double Open Dirt Drilling Bucket, shown in Fig. 1, was used to complete the shaft excavation to its' final depth of 32.6 m (106.9 ft). Polymer slurry was used to maintain an open excavation through the uncased zone. Bottom cleaning was performed with this drilling bucket after the auger teeth plate was removed forming a flat bottom cleanout bucket. After cleanout, a full-length, 2,083 mm O.D (82 in) reinforcing cage was installed in the shaft. Concrete, with a design slump of 203 mm (8 in) was then pumped and placed in the shaft via a 254 mm I.D. (10 in) tremie pipe. The final shaft base was at Elevation −32.01 m (−105.0 ft) and the final top of shaft concrete was at Elevation +1.19 m (+3.9 ft).

Fig. 1. Double open drilling bucket.

3 Test Shaft QA/QC

3.1 Shaft Area Profile Evaluator

A SHaft Area Profile Evaluator or SHAPE device was used to determine the characteristics of the wet pour, drilled shaft excavation. The device was pin connected to the drill rig Kelly bar and then lowered into and retrieved from the shaft excavation. While in the wet excavation, this device used eight ultra-sonic pulsers to scan the side walls at a rate of approximately one scan per second. It provided a quick check of the excavation verticality, radii, shape, and drilled hole volume. The test was performed a few hours before inserting the reinforcing cage in the hole. A photograph of the device prior to insertion into the test shaft is provided in Fig. 2. Test results, presented in Figs. 9 and 10, are discussed later in this document.

Fig. 2. Shaft area profile evaluator.

3.2 Shaft Quantitative Inspection Device

A Shaft QUantitative Inspection Device or SQUID was used to check the base cleanliness two hours prior to placing the reinforcing cage and initiating the concrete pour. This device used three 10 cm^2 (1.55 in^2) cone penetrometers and three 521 mm (6 in) displacement plates to assess base cleanliness. For the purpose of sediment thickness measurements, flat tips were attached to the penetrometers for the testing. A photograph of the device prior to lowering it into the test shaft is provided in Fig. 3. Base cleanliness tests were performed in the center of the shaft as well as in North, East, and West quadrants of the shaft. The South quadrant test was not performed due to limitations on the drill rig reach. The SQUID was pushed into the base material by the weight of the drill rig Kelly bar. The device penetrometers can be pushed to a maximum penetrometer pressure of 100 MPa (14 ksi). Test results are presented in Figs. 11 and 12, with test result discussion provided later in this paper.

Fig. 3. Shaft quantitative inspection device.

3.3 Bi-Directional Static Load Testing

Three 6.7 MN (750 ton) capacity GRL-Cells were used in the bi-directional jack assembly. This multi-cell jack assembly was located 4.2 m (13.8 ft) above the shaft base elevation. The jack-assembly was capable of producing a 20 MN (2,250 ton) jack load and a maximum bi-directional test load of 40 MN (4,500 tons). A photograph of the bi-directional jack assembly prior to insertion in the test shaft is provided in Fig. 4. Bi-directional test results are presented in Figs. 13, 14, and 15 with discussion presented later in this paper.

3.4 Cross-hole Sonic Logging

Access tubes for cross-hole sonic logging were attached to the reinforcing cage at 300 mm (12 in) spacings around the interior of the reinforcing cage. The 38 mm (1.5 in) O.D. steel access tubes ran the full length of the cage, and were pre-cut at the location of the bi-directional jack assembly. The cut access tubes were fitted with expandable couplers at the jack assembly location to accommodate jack expansion. Figure 5 contains a photograph of cross-hole sonic logging being performed on the test shaft. Four profiles were collected and processed concurrently with the CHAMP-Q cross-hole sonic logging system. The multiple transceivers were pulled upwards from the bottom of the CSL access tubes with ultrasonic signals acquired at 50 mm (2 in) vertical intervals as the four transceivers were concurrently raised.

Fig. 4. 20 MN bi-directional jack assembly attached to the reinforcing cage.

The steel cross-hole sonic logging access tubes are shown attached to the interior of the reinforcing cage in Fig. 6. The expandable couplers used on the CSL tubes at the location of the bi-directional cells are visible in the bi-directional jack assembly photograph in Fig. 4. CSL test results are presented in Figs. 16 and 17, with discussion of the test results provided in a subsequent section of this paper.

3.5 Thermal Integrity Profiling

Thermal Integrity Profiling cables were attached to the reinforcing cage at 300 mm (12 in) spacings around the interior of the reinforcing cage. Each individual thermal wire cable had thermal sensors spaced 300 mm (12 in) apart along the length of the cable. The thermal wire cables ran the full length of the reinforcing cage with slack provided at the bi-directional jack assembly location. Figure 6 illustrates the thermal wire being attached to the reinforcing cage.

Immediately after completion of the concrete pour, one Thermal Aggregator (TAG) and seven Thermal Acquisition Ports (TAP-Edge) data logging units were attached to the thermal wire cables. Every 15 min, the temperature of each thermal sensor was read by the data loggers. The collected thermal readings from each wire

Fig. 5. Cross-hole sonic logging test using multiple transceivers.

Fig. 6. Thermal integrity profiling wire attachment.

were pushed to the Cloud where they could be viewed in real time. The shaft reached its' peak temperature approximately 26.3 h after the start of the concrete pour. The thermal integrity profiling results are presented in Figs. 19, 20, and 21, with test results discussed later in this paper.

3.6 Concrete Placement

During concrete placement, measurements to the top of concrete in the excavation along with the volume of concrete placed were recorded. The concrete volume in both the tremie pipe and the pump lines was subtracted from the concrete volume placed.

4 QA/QC Test Results

4.1 Excavated Volume

The volume of the shaft excavation has historically been determined post-pour based on the amount of concrete placed in the shaft. The placed concrete volume can be plotted versus depth by measuring the depth to the top of concrete following discharge of each concrete truck and its associated concrete volume from batch plant tickets. The plot of concrete volume versus depth is routinely compared to the theoretical volume versus depth to assess potential integrity issues. A sudden increase in concrete volume compared to the theoretical volume at a given depth is indicative of an oversized shaft due to a bulge or cavity filling. Conversely, a sudden decrease in concrete volume compared to the theoretical volume at a given depth is indicative of an undersized shaft due to an inclusion or necking.

Figure 7 presents a plot of the placed concrete volume versus depth. This figure also includes the theoretical shaft excavation volume versus depth. SHAPE results of the excavated hole volume are also included in this figure. The SHAPE results closely mirror the theoretical volume which appears reasonable given that 58% of the shaft length was permanently cased. Construction concreting records indicate the placed volume was 104% of the theoretical volume. Based on the shaft construction techniques, this implies the extra concrete volume should be located in the lower uncased portion of the shaft.

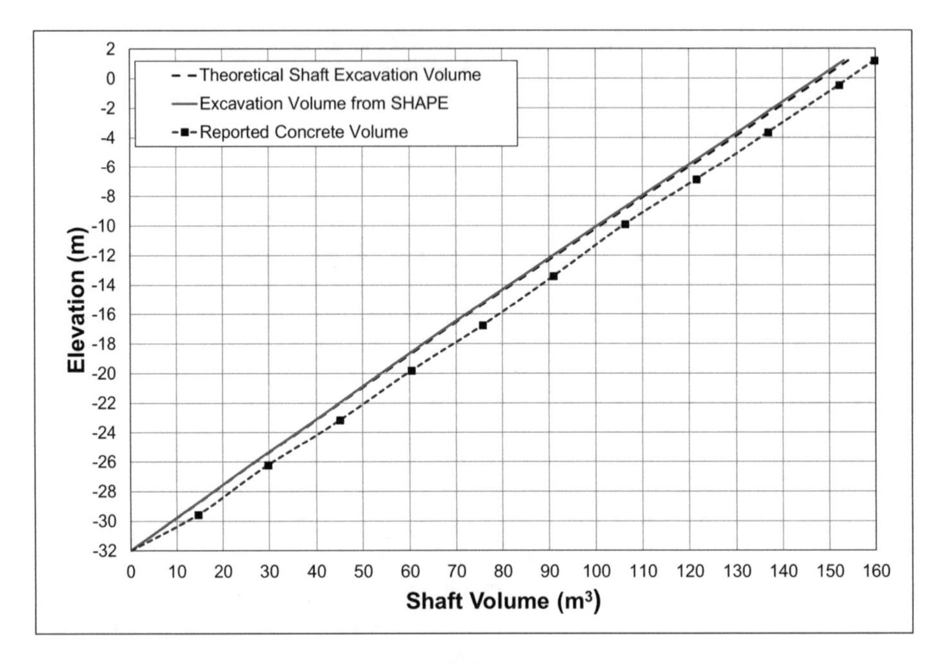

Fig. 7. Comparison of shaft volumes versus elevation.

4.2 Shaft Diameter

The average radius and diameter of the constructed shaft was calculated from three methods; the placed concrete volume versus depth, the SHAPE results, and the thermal integrity profiling results. A comparison of the average shaft diameter from these methods as well as with the theoretical shaft diameter versus depth is presented in Fig. 8.

The concrete volume information yields the largest shaft diameter below Elevation −22 m (−72 ft) in the uncased portion of the shaft. The concrete volume information also yields the largest shaft diameter above Elevation −4 m (−13 ft) in the permanently cased portion of the shaft which is unreasonable. The shaft diameter calculated by thermal integrity profiling clearly shows the bi-direction jack assembly at Elevation −27.7 m (91.0 ft) due to the heat sink associated with the steel cells at that location. Above Elevation −10.0 m (32.8 ft), the shaft diameter indicated by the thermal integrity profiling is slightly greater than the diameter of the permanent casing. The shaft diameter calculated by SHAPE appears to be the most realistic representation of the constructed shaft.

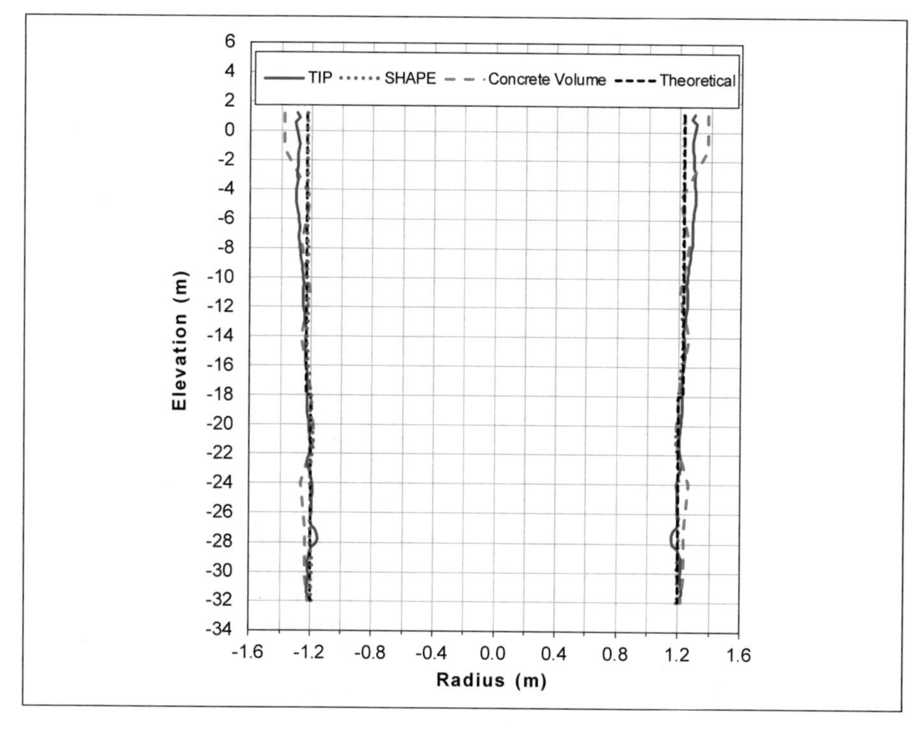

Fig. 8. Comparison of average shaft radius versus elevation.

4.3 Verticality

The project specifications required the shaft verticality to not deviate from the plan alignment by more than 6.4 mm (¼ in) per 304.8 mm (1 ft) of depth or 2%. Hence, for the 33.2 m (108.9 ft) long test shaft, the maximum allowable deviation would be 0.69 m (2.26 ft) at the base. Figure 9 presents the SHAPE scan results from the each of four profiles. Note that the depth scale in the scan results is referenced to the top of the drilling slurry at Elevation +3.05 m (+10.0 ft).

As indicated in the profile schematic, sensor 1 is positioned to the north to conduct the test. The four scan profiles indicate the drilled shaft base drifts slightly towards the west. Figure 10 presents the SHAPE determined verticality and eccentricity. The verticality of 0.20% based on a base eccentricity of 0.07 m (0.23 ft) is well within the above noted specification limits. The lower portion of Fig. 10 also displays the calculated encroachment area of 0.16 m^2 (1.72 ft^2) into the shaft sidewall.

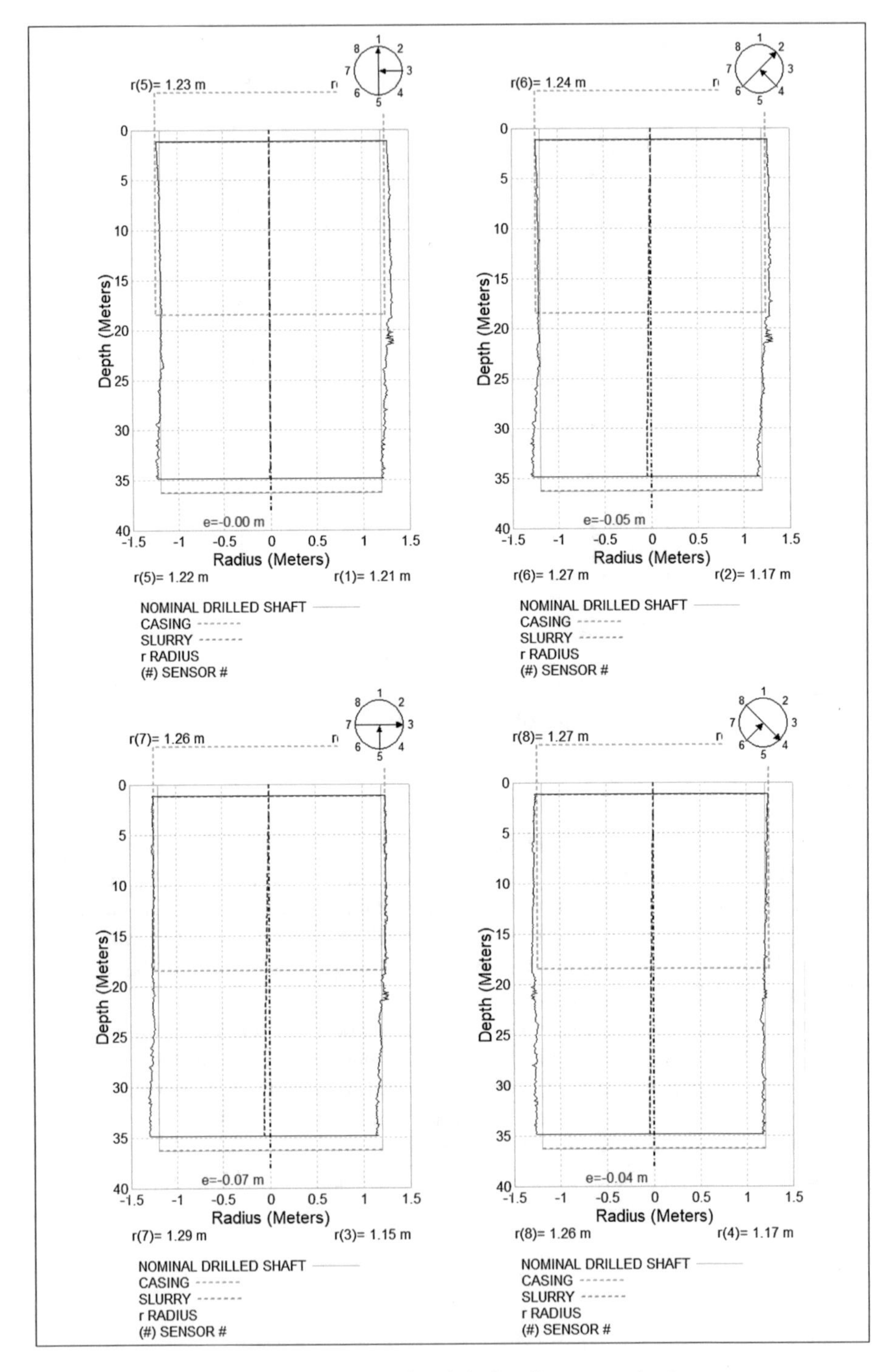

Fig. 9. SHAPE results of shaft radius versus depth.

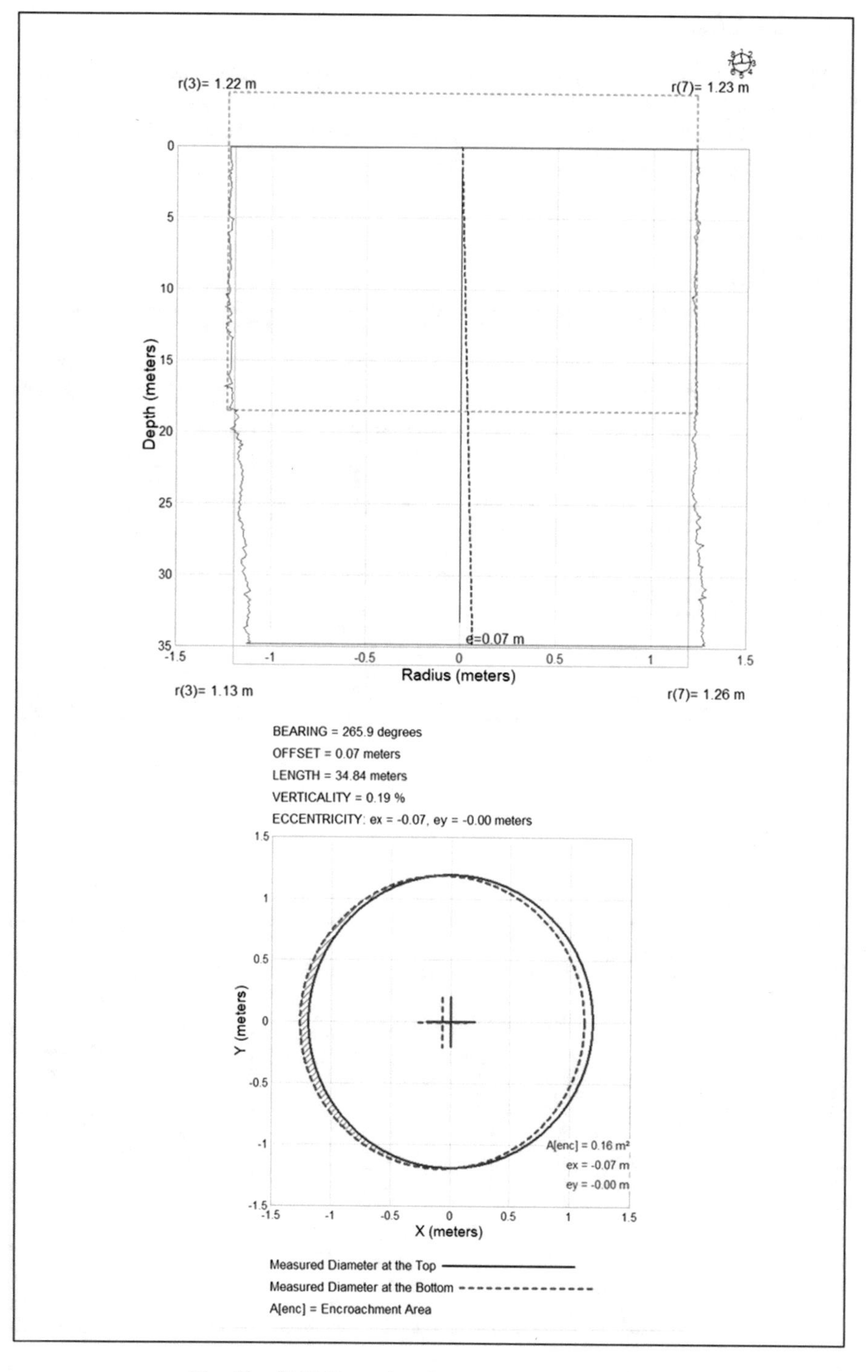

Fig. 10. SHAPE results of shaft radius versus depth.

4.4 Base Cleanliness

The project specifications stipulated that less than 13 mm (0.5 in) of sediment or debris be present over a maximum of 50% of the shaft base area at the time of concrete placement. In addition, the maximum sediment or debris present at any location on the base was required to be less than 38 mm (1.5 in).

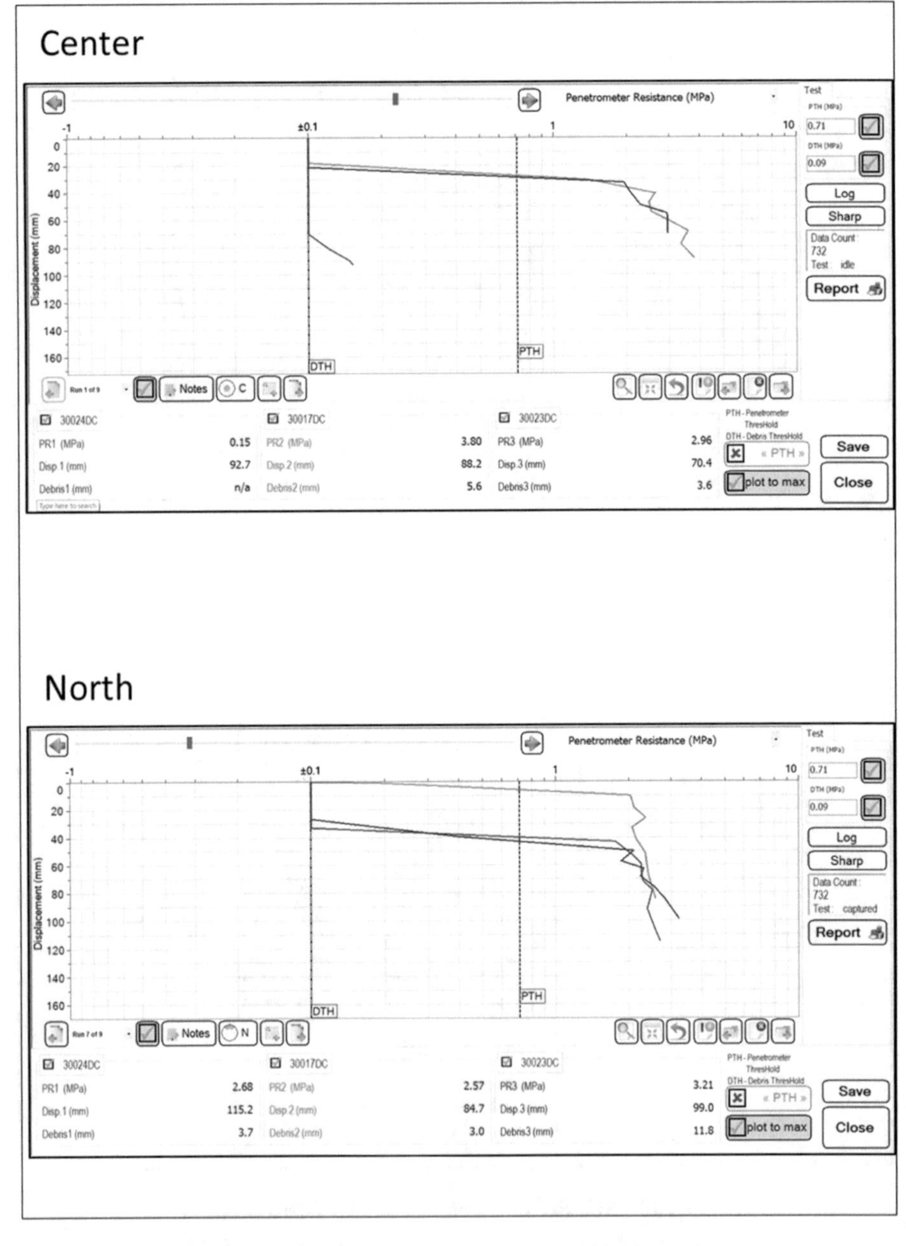

Fig. 11. SQUID results at shaft center and North quadrant.

Figures 11 and 12 present the SQUID test results from four test locations; Center, North, East, and West. Since the primary purpose of the tests were to evaluate base cleanliness, the penetrometers were fitted with a flat tip rather than the conventional cone tip. Base cleanliness was assessed according to the criteria proposed by Moghaddam et al. (2018). The first vertical line, labeled DTH, is associated with penetration resistance associated with debris. Values less than DTH are associated with very soft materials that will be readily displaced or due to an uneven base condition causing a debris plate to hang atop a grooved or uneven surface. The second vertical

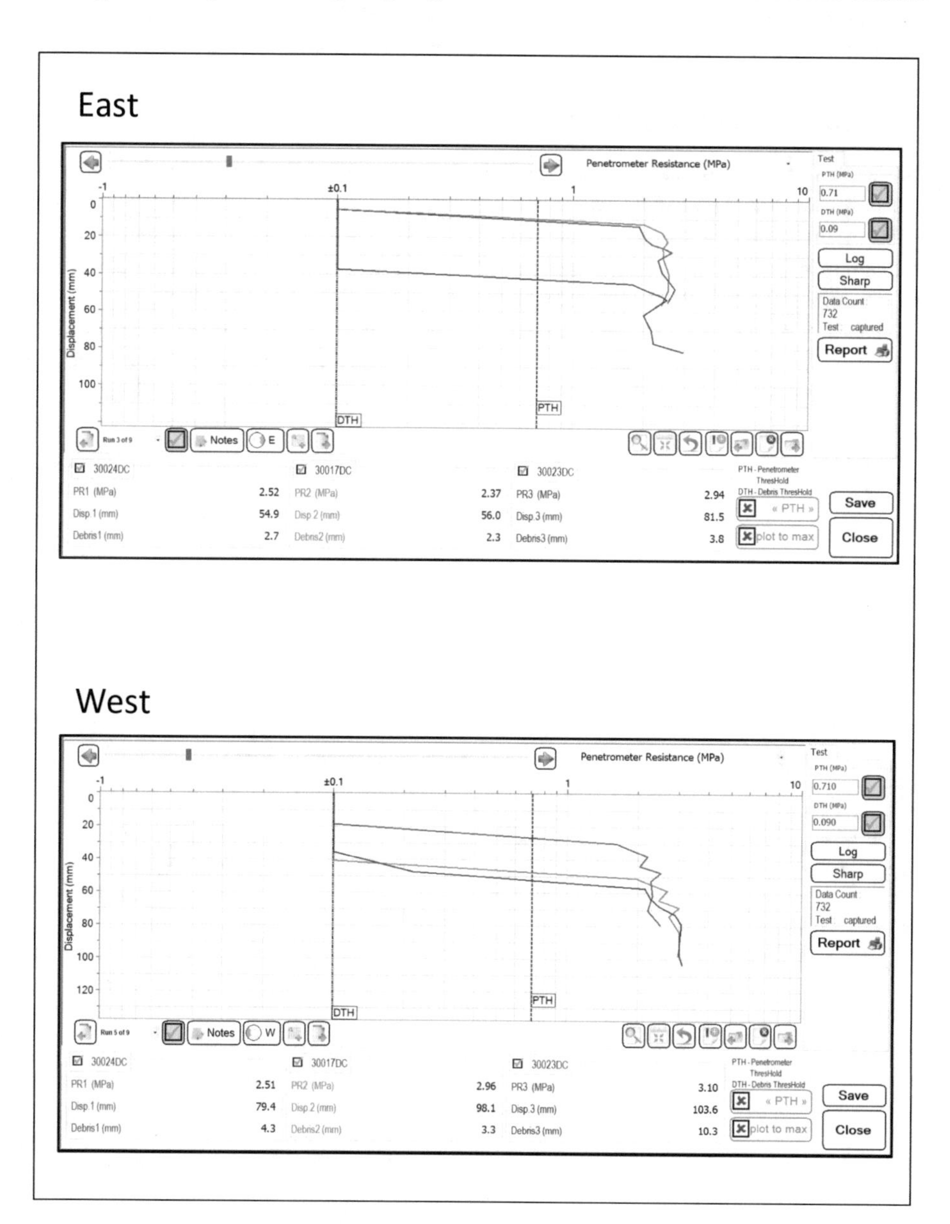

Fig. 12. SQUID results at East and West quadrants.

line, labeled PTH, corresponds to the penetration resistance offered by natural soils. The measured displacement between crossing the DTH and PTH thresholds is the defined debris thickness.

The test results in Figs. 11 and 12 indicate the debris thickness is typically on the order of 5 mm (0.2 in) or less across the test locations. Two of the eleven penetrometer locations had slightly greater debris thicknesses of 11.8 and 10.3 mm (0.5 and 0.4 in). However, the 13 mm (0.5 in) debris limit was not exceeded at any location. Therefore, less than 50% of the tested shaft base area had less than 13 mm (0.5 in.) of debris and no reading indicated more than the maximum allowed debris thickness at any location of 38 mm (1.5 in). Hence, the test shaft base was very clean.

4.5 Geotechnical Design and Capacity

Based on the geotechnical design calculations, the test shaft had an estimated nominal resistance of 20.6 MN (4630 kips). The test shaft was expected to carry approximately 71% of this nominal resistance or 14.6 MN (3287 kips) in shaft resistance and the remaining 6.0 MN (1348 kips) at the shaft base. The pretest unit base resistance was therefore anticipated by the foundation designer to be 1.34 MPa (28 ksf).

BDSLT results are presented in Fig. 13 with the resulting internal force profiles presented in Fig. 14. As noted earlier, the test shaft had an anticipated nominal resistance of 20.6 MN (4630 kips) with 14.6 MN (3287 kips) in shaft resistance and

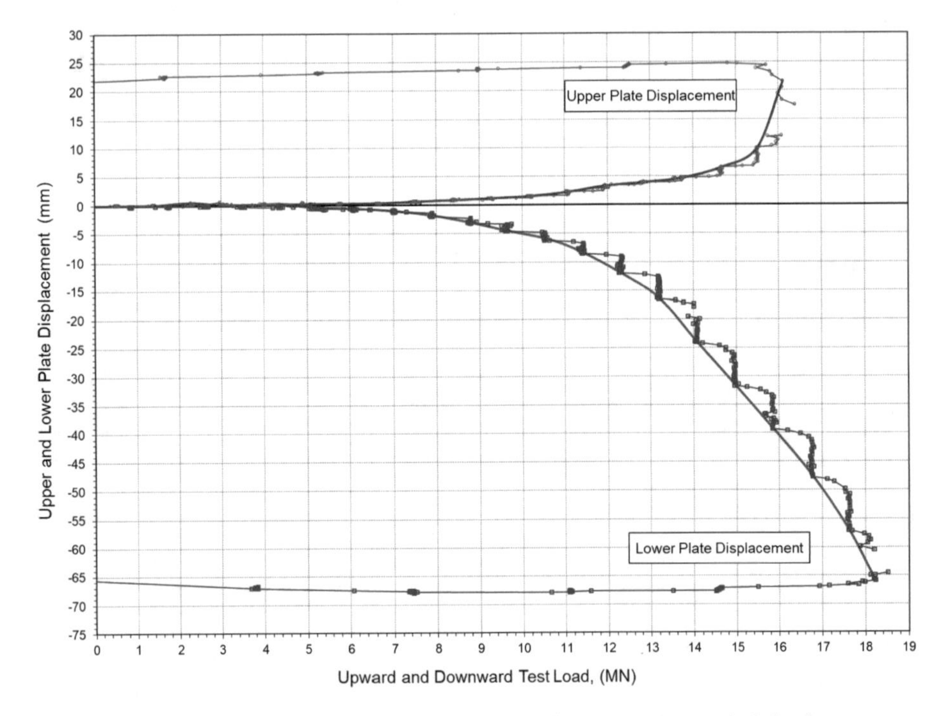

Fig. 13. BDSLT upper and lower bearing plate movement vs jack load.

6.0 MN (1348 kips) at the shaft base. The BDSLT results indicate the test shaft had a substantially greater nominal resistance of 34.4 MN (7733 kips) with 26.2 MN (5890 kips) of shaft resistance and 8.2 MN (1843 kips) of base resistance. Hence, test results confirmed the test shaft met the foundation design requirements and further design optimization could be achieved by shortening production shaft lengths.

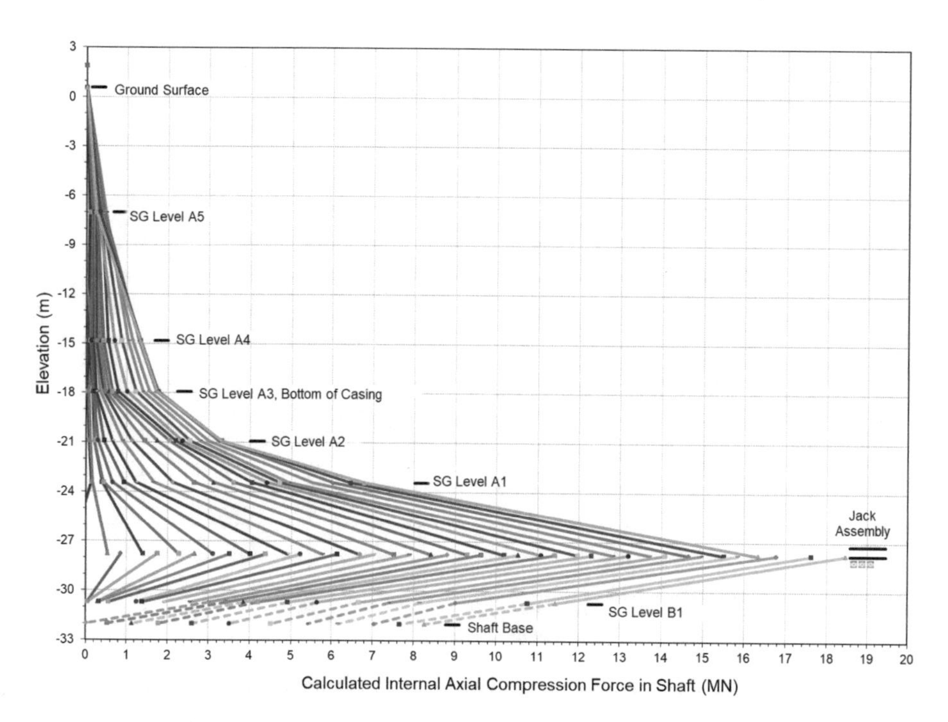

Fig. 14. BDSLT determined internal force profile vs elevation.

4.6 Unit Base Resistance

The FHWA drilled shaft design manual, Brown et al. (2018), also includes a design procedure for unit base resistance in cohesionless materials using Standard Penetration Test (SPT) N_{60} values. The SPT N_{60} value within 2 diameters below the shaft base elevation ranged from 10 to 20 and had an average value of 13. According to the SPT design procedure in the FHWA manual, the unit end bearing resistance would be 0.75 MPa (15.6 ksf) at a displacement of 5% of the base diameter. The unit base resistance from this design method is less than the geotechnical designer's anticipated unit base resistance.

The unit base resistance was calculated from the bi-directional static load test (BDSLT) result internal force profile in Fig. 14 by assuming the foundation segment between the lowest strain gage level and the shaft base had the same unit shaft resistance as the overlying segment. This calculated unit base resistance was greater than both the geotechnical designer's method and the FHWA SPT method unit base resistances noted above. It should be noted that a base displacement of 119 mm (4.7 in) corresponding to 5% of the base diameter was not achieved in the bi-directional static load test. The maximum unit base resistance determined from the BDSLT, presented in Fig. 15, was 1.77 MPa (37 ksf) at a displacement of 2.5% of the base diameter.

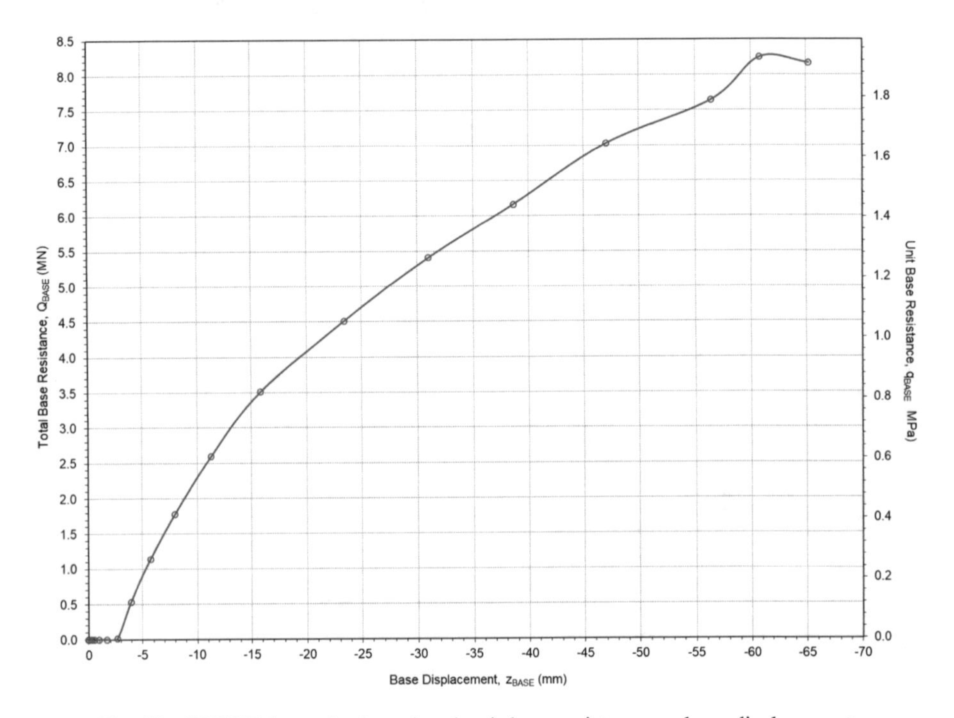

Fig. 15. BDSLT determined total and unit base resistance vs base displacement.

SQUID penetrometers have a maximum penetration distance into the base material of 150 mm (6 in.). Even with this limited penetration depth, the penetrometer force versus displacement results indicated geotechnical failure occurred between 1.8 to 2.1 MPa (37 to 44 ksf). Geotechnical failure was defined as the break point in the unit resistance versus displacement plots in Figs. 11 and 12. In the base materials at this site, the break point was within 20% of the unit end bearing result from the bi-directional test. Obviously, the zone of influence of the 10 cm^2 (1.44 in^2) penetrometer and the 2.4 m (94 in) diameter shaft are significantly different. Therefore, unit base resistance correlations are anticipated to be meaningful only when a uniform material exists within the zone of influence beneath the shaft base.

4.7 Concrete Quality and Cover

The test shaft was cast with eight CSL access tubes uniformly spaced around the interior of the reinforcing cage. This resulted in a total of 28 possible CSL profiles from the various tube combinations. A representative selection of CSL profiles is presented in Figs. 16 and 17. Each ultrasonic CSL profile consists of two graphs. The left-hand

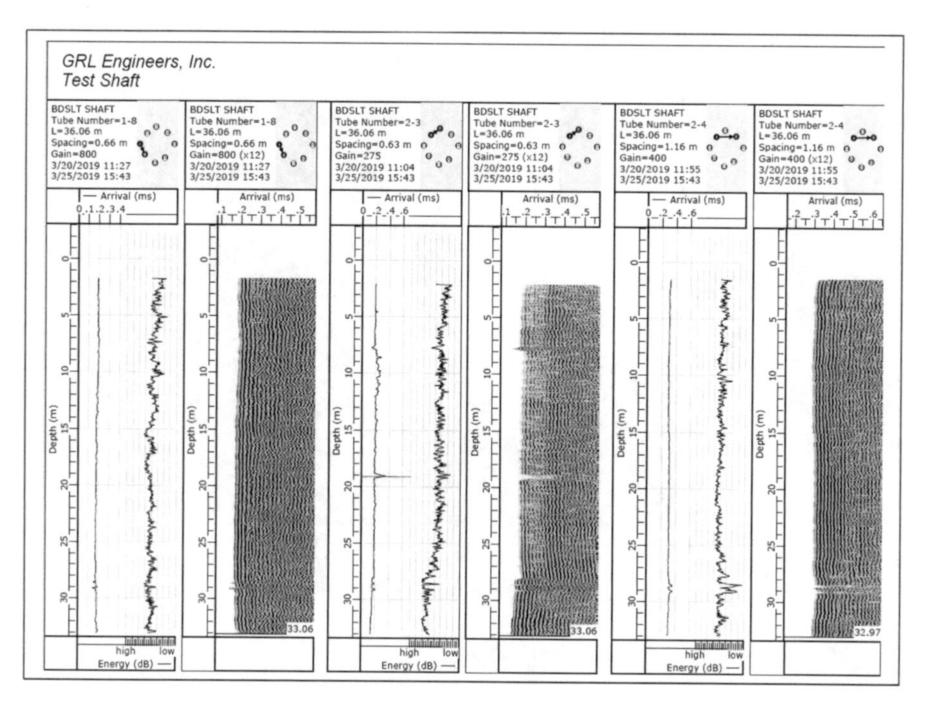

Fig. 16. Representative CSL profiles 1–8, 2–3, 2–4.

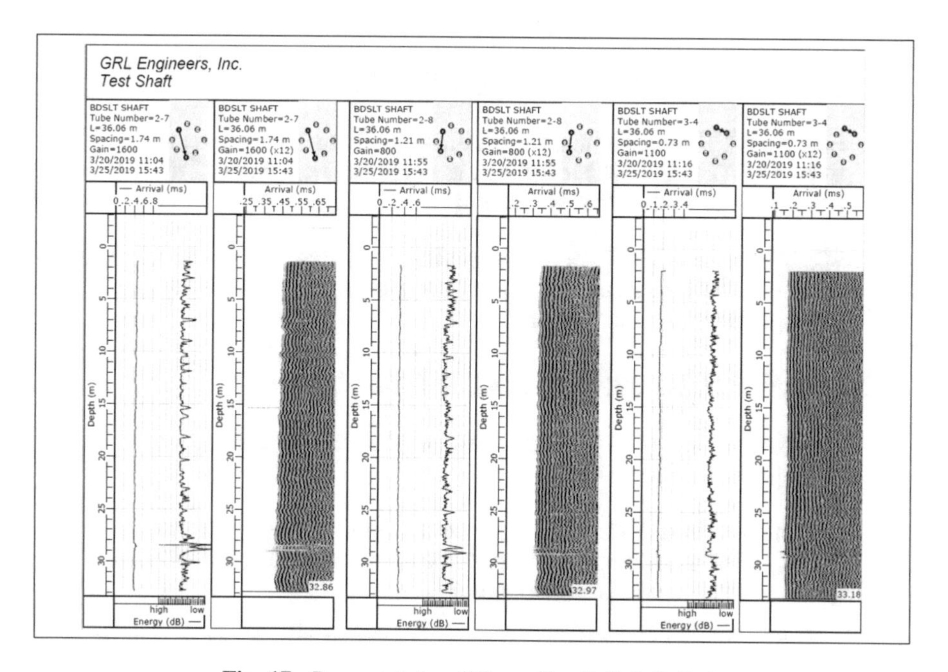

Fig. 17. Representative CSL profiles 2–7, 2–8, 3–4.

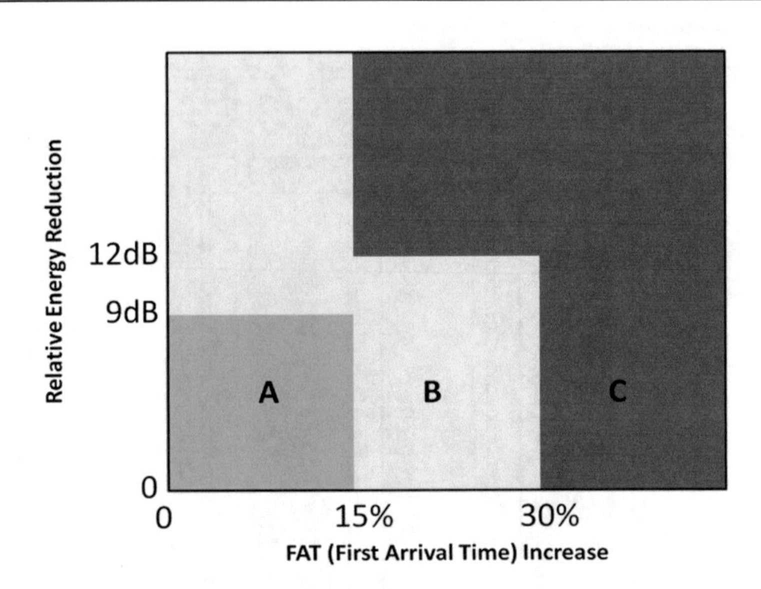

Class A: Acceptable CSL test results

- First Arrival Time (FAT) increases are less than 15% of the local average FAT value AND reductions in energy are less than 9 dB of the local average value of relative energy.

Class B: Conditionally Acceptable CSL test results

- First Arrival Time (FAT) increases between 15 and 30% of the local average FAT value AND reductions in energy are less than 12 dB of the local average value of relative energy.

or

- First Arrival Time (FAT) increases are less than 15% of the local average FAT value AND reductions in energy are greater than 9 dB of the local average value of relative energy.

Class B: Highly Abnormal CSL test results

- First Arrival Time (FAT) increases are greater than 30% of the local average FAT value.

or

- First Arrival Time (FAT) increases are greater than 15%% of the local average FAT value AND reductions in energy are greater than 12 dB of the local average value of relative energy.

Fig. 18. Proposed CSL rating criteria (after DFI, 2019).

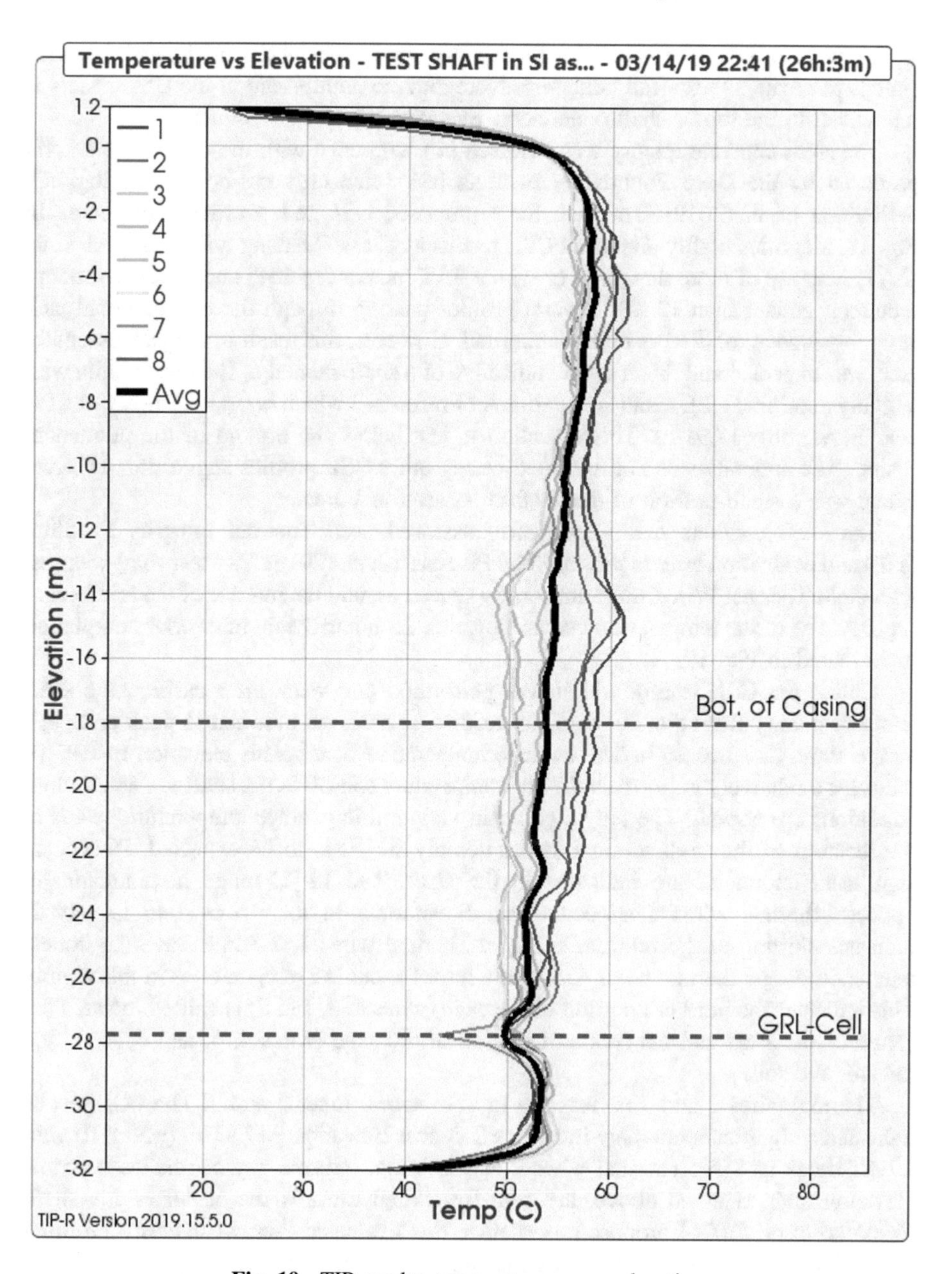

Fig. 19. TIP results: temperature versus elevation.

graph for a given profile presents a plot of the first arrival time of the received ultrasonic signal at each sampling depth. The left-hand plot also displays the calculated relative energy of the received signal at each sampling depth. Delays in the first arrival time and/or decreases in the received signal energy are indicative of anomalies in the shaft concrete. The right-hand plot for a given profile stacks the modulated raw data signals to create a "waterfall" diagram. Note that the depth scale in the CSL results is referenced to the top of shaft concrete at Elevation +1.19 m (+3.9 ft).

The shaft concrete quality was assessed in accordance with the evaluation criteria proposed by the Deep Foundation Institute task force on cross-hole sonic logging, Sellountou et al. (2019). The task force proposed CSL rating criteria, presented in Fig. 18, identifies highly abnormal CSL results as Class C having a First Arrival Time (FAT) increase of more than 30% or with a FAT increase of 15% and a relative energy reduction greater than 12 dB. Several profiles passing through the bi-directional jack assembly were classified as highly abnormal. However, this result in the jack assembly zone was expected and this it is not indicative of a shaft anomaly. The only profile with a highly abnormal CSL result was perimeter Profile 2–3 which had a FAT delay of 81% near Elevation −17.91 m. This elevation is just below the bottom of the permanent casing. The anomaly was not indicated in any other CSL profiles suggesting its' areal extent was a small portion of the shaft cross sectional area.

The shaft concrete quality was also assessed used Thermal Integrity Profiling (TIP) and evaluation criteria proposed by Piscsalko et al. (2016). The test shaft was cast with eight Thermal Wire Cables uniformly spaced around the interior of the reinforcing cage. A plot of the temperature versus elevation 26 h and 3 min after pour completion is presented in Fig. 19.

Unlike the CSL testing which was performed one week after casting, the shaft concrete quality from the TIP data can readily be assessed between ½ peak and peak temperature, or 13 to 26 h. The average temperature data versus elevation in Fig. 19 indicates a relatively uniform shaft. The temperature roll-off at the shaft top and bottom conditions are normal. The only significant variation in average temperature occurs at the location of the bi-directional jack assembly which is to be expected. Hence, no significant anomalies are indicated in the shaft. The 12 °C range in diametrically opposite thermal wires that occurs near Elevation −14 to −15 m (−46 to −49 ft) indicates shifting of the reinforcing cage. Thermal wires 1, 2 and 8 have the hottest temperatures and thermal wires 4, 5, and 6 have the coolest temperatures in this region. This indicates the northern portion of the cage (wires 1, 2, and 8) is shifted towards the center of the shaft and the southern portion of the cage (wires 4, 5, and 6) is shifted towards the soil.

Thermal wires 2 and 3 correspond to CSL access tubes 2 and 3. The CSL results indicated a significant anomaly in Profile 2–3 near Elevation −17.91 m (−58.8 ft) with a FAT delay of 81%. Thermal wires 2 and 3 do not indicate any anomalies near this elevation and, as noted above, are actually shifted towards the center of the shaft. Piscsalko et al. (2016) proposed a criterion that evaluated thermal integrity profiling results based on the structural and geotechnical impact of an anomaly. According to this criterion, a shaft was considered satisfactory if the effective average radius reduction was 0 to 6% and the local cover criteria was met. An anomaly was indicated if the thermal integrity profiling results had an effective average radius reduction greater than 6% and the local cover criteria was not met. Based on this criterion, the test shaft would be acceptable as the effective average radius reduction did not exceed 6% and a concrete cover criterion was not specified.

A plot of the radius and concrete cover versus elevation is presented in Fig. 20. The calculated average shaft radius is greater than required shaft radius throughout except were affected by the GRL-Cells. Similarly, the average concrete cover is 150 mm (3 in) or greater. Note that the cage shifting previously described results in a concrete cover on the order of 100 mm (2 in) in the vicinity of wires 5 and 6 from Elevation −14 to −26 m (−46 to −85 ft). A three-dimensional depiction of the shaft with an overlay of the reinforcing cage is presented in Fig. 21. This figures also includes a generalized soil profile.

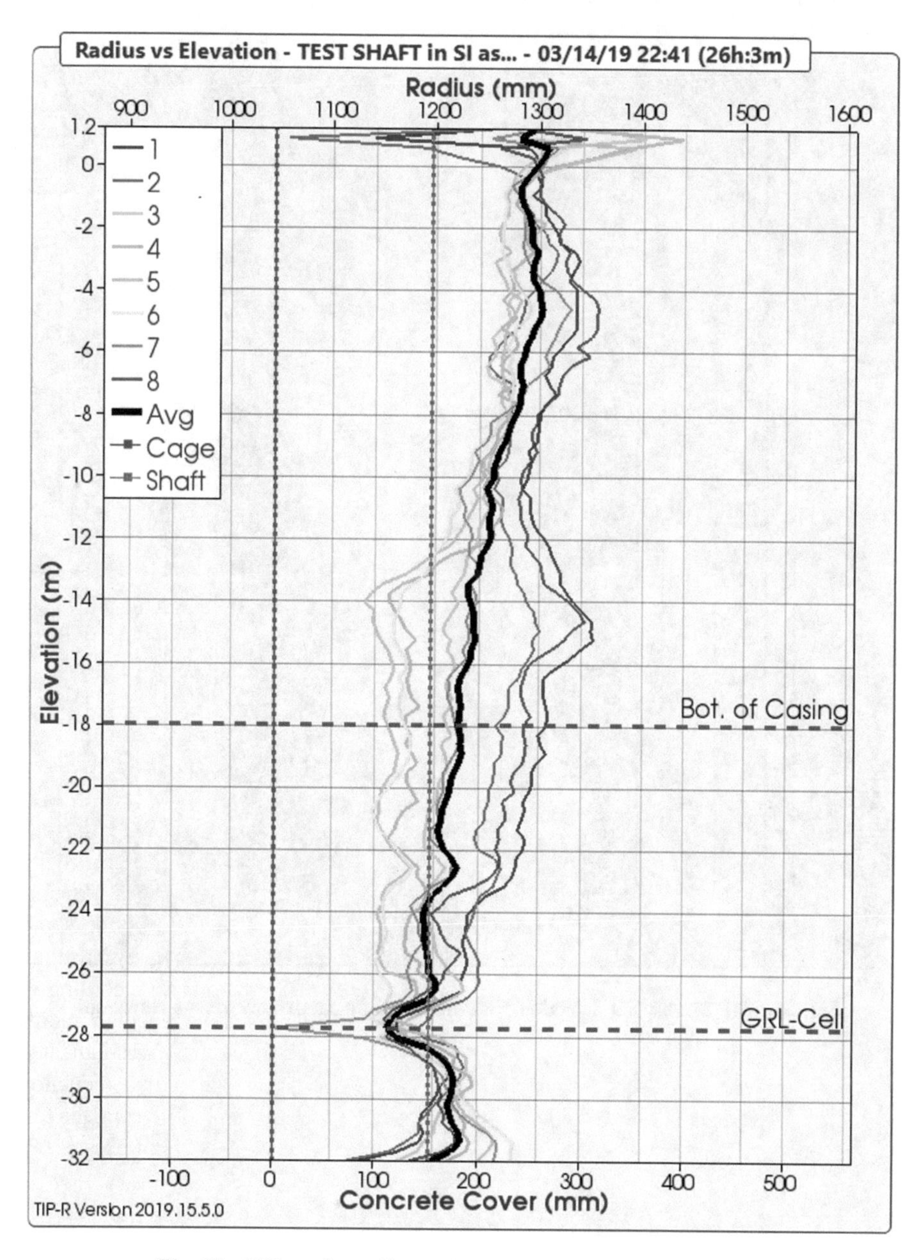

Fig. 20. TIP results: radius and concrete cover versus elevation.

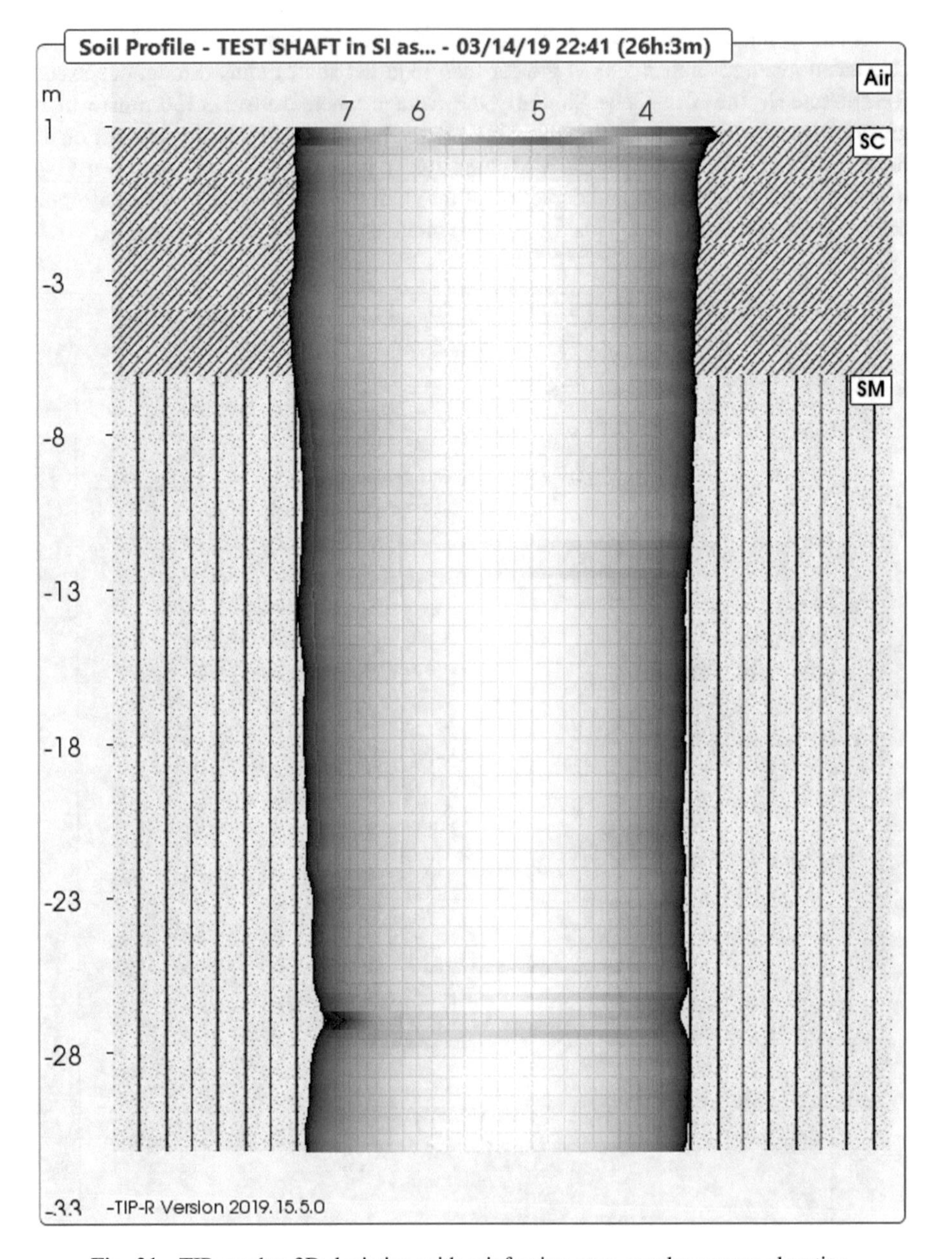

Fig. 21. TIP results: 3D depiction with reinforcing cage overlay versus elevation.

5 Conclusions

Several benefits were obtained from the installation and QA/QC testing of a test shaft on the project prior to production shaft installation. The test shaft QA/QC tests confirmed that the contractor's means and methods of shaft installation resulted in a shaft meeting verticality, base cleanliness, concrete quality and integrity requirements. The bi-directional static load test confirmed that the design was achievable and that further optimization of shaft lengths were possible.

The SQUID testing of the shaft base condition indicated the presence of minimal debris and a very clean shaft base. Bi-directional static load test results indicated a greater unit base resistance than used in the design as well as that indicated by a frequently used design method. A promising correlation between SQUID penetration resistance and unit base resistance was also obtained for the base materials tested.

The shaft verticality and base cleanliness were quickly checked by the SHAPE and SQUID equipment, respectively prior to shaft concrete placement. These devices showed the benefit of newer technology in shaft construction and quality control.

The shaft concrete quality of the 2.38 m (7.83 ft) diameter drilled shaft was evaluated by cross-hole sonic logging (CSL) seven days after concrete placement and by thermal integrity profiling (TIP) one day after concrete placement. Both methods indicated a high-quality shaft. The localized highly abnormal CSL result was not apparent in the corresponding TIP results. The anomaly was therefore considered insignificant based on the thermal integrity profiling results. A significant reduction in the time required for shaft acceptance is possible with TIP as the test results were available 6 days earlier than the CSL results.

References

AASHTO LRFD Bridge Construction Specifications, 4th Edition: American Association of State Highway and transportation Officials, Washington, D.C., 574 p. (2017)

Brown, D.A., Turner, J.P., Castelli, R.J., Loehr, E.L.: Drilled Shafts: Construction Procedures and Design methods, Geotechnical Engineering Circular No. 10, FHWA Report No. FHWA-NHI-18-024, National Highway Institute, U.S. Department of Transportation, Federal Highway Administration, Washington DC, 754 p. (2018)

Terminology and Evaluation Criteria of Cross-hole Sonic Logging (CSL) as applied to Deep Foundations, Deep Foundations Institute Task Force, Deep Foundations Institute (2019)

Moghaddam, R., Hannigan, P., Anderson, K.: Quantitative assessment of drilled shafts base-cleanliness using the shaft quantitative inspection device (SQUID). In: International Foundations Congress and Equipment Exposition, IFCEE 2018, Orlando, FL (2018)

Piscsalko, G., Likins, G., Mullins, G.: Drilled shaft acceptance criteria based upon thermal integrity. In: DFI 41st Annual Conference on Deep Foundations, New York, NY, pp. 1–10. Deep Foundations Institute (2016)

A Case Study on Buckling Stability of Piles in Liquefiable Ground for a Coal-Fired Power Station in Indonesia

Muhammad Hamzah Fansuri[1(✉)], Muhsiung Chang[1], and Rini Kusumawardani[2]

[1] Department of Civil and Construction Engineering, National Yunlin University of Science and Tech (YunTech), Yunlin, Taiwan
m10616212@gemail.yuntech.edu.tw, changmh@yuntech.edu.tw

[2] Department of Civil Engineering, Universitas Negeri Semarang (Unnes), Semarang, Indonesia
rini.kusumawardani@mail.unnes.ac.id

Abstract. This paper discusses a case study on the assessment of buckling stability of piles due to liquefaction of foundation soils for a coal-fired power station (CFPS) in Indonesia. As the rapid growth in the economy sector, the demand for electricity is increasing and a CFPS is planned and constructed in Central Java. During the planning stage, the risk caused by earthquakes should be considered. As the foundation soils of the site consist of soft sandy silts or clays interbedded with loose fine sands up to a depth of 9 m, soil liquefaction and its effect on the buckling stability of piles for CFPS thus become the main concerns of this project. Liquefaction analysis is performed based on a SPT-N approach. A depth-weighted procedure is applied for the assessment of liquefaction potential for the site. Results of liquefaction assessment indicate the site is prone to risk of soil liquefaction due to the design earthquake. Buckling of piles due to seismic loading is evaluated for the cases of liquefied soils in both dry and wet seasons, while only the wet season scenario is the main focus of this paper. A buckling stability index $\boldsymbol{G}$ is adopted as the difference between the critical pile length ($\boldsymbol{H_c}$) for buckling and the unsupported pile length ($\boldsymbol{D_L}$) due to soil liquefaction. If $\boldsymbol{G}$ is greater than zero, then the pile is safe; otherwise, the pile will buckle. Results of buckling assessment show $\boldsymbol{G} > 0$ for the piles of the site with an average $\boldsymbol{G}$ value of 15 as the foundation soils are liquefied during the design earthquake with magnitude $\boldsymbol{M_w}$ of 6.8, indicating the pile foundation of the CFPS should be safe from buckling failure due to soil liquefaction.

1 Introduction

Indonesia is one of the countries located in a highly seismic area. It is surrounded by the Trans-Asian and Circum Pacific belts. In addition, it is also surrounded by three major active tectonic plates, namely, Eurasia, Indo-Australian, and Philippine Plates. Due to the effect of the colliding plates, the region becomes actively tectonic and volcanic as well. The active tectonics are related to the same hazards or disasters such as earthquake, tsunami, fault, uplift, subsidence and mass-movement. In the last decade

H. El-Naggar et al. (Eds.): GeoMEast 2019, SUCI, pp. 88–106, 2020.
https://doi.org/10.1007/978-3-030-34252-4_8

major earthquakes occurred in several parts of the islands of Indonesia causing damages and many fatalities.

Indonesia has often been hit by huge disasters, such as earthquake and liquefaction. Earthquakes with magnitude (M_w) of 5.9 (2006), 7.6 (2009), 7 (2018) induced sand boiling and ground settlements in Yogyakarta, Padang, North of Lombok, respectively (Unjianto 2006; Tohari 2013; Kasbani 2018). Recently, the Palu earthquake occurred on 28 September 2018 with magnitude (M_w) of 7.4 caused strong shaking, generating a tsunami and massive liquefaction (GEER 2019). Therefore, increased attention should be focused on earthquakes and subsequent liquefaction potential in Indonesia.

In order to reduce the damage caused by an earthquake, the engineering profession needs to take into account the risks caused by seismic loading. Seismic loading is triggered by the earthquake vibration waves on the soil layer. Thus, it affects the soil behavior in order to support the structure, both structures beneath the soil or above.

Certain soils liquefy during earthquake shaking, losing its shear strength causing it to flow taking with it any overlying non-liquefied crust. As illustrated in Fig. 1. These soil layers drag the pile with them causing bending failure. In terms of soil-pile interaction, this mechanism assumes that the soil pushed the pile (Hamada and O'Rourke 1992; Finn and Fujita 2002). Soils that can liquefy are clearly understood and can be identified by evaluation procedures proposed by Youd et al. (2001). See "liquefaction assessment at foundation soil" further in this report.

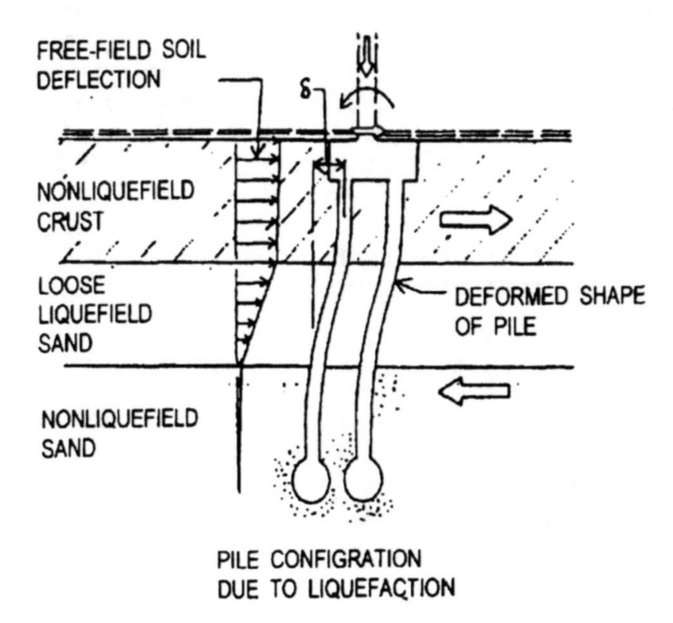

Fig. 1. Potential failure mode of piles due to seismic loading and soil liquefaction (Finn and Thavaraj 2001)

Foundations directly supported on soil are particularly vulnerable to liquefaction. This phenomenon is well understood and studied. Piled foundations and their response to liquefaction are less studied.

In structure terms, piles are slender columns with lateral support from the surrounding soil. Generally, as the length of the pile increases, the allowable load on the pile increases due to the additional shaft friction but the buckling load decreases inversely with the square of the length. If unsupported, these columns will fail due to buckling instability and not due to the crushing of the material. During earthquake-induced liquefaction, the soil surrounding the pile loses its effective confining stress and perhaps not offers sufficient lateral support. Hence, the pile now acts as an unsupported length column prone to axial instability. The instability may trigger the pile to buckle sideways in the direction of least elastic bending stiffness under the axial action load. In this case, the pile could push the soil and it could not be necessary to invoke lateral spreading of the soil to cause a pile to collapse (Bhattacharya 2003). If the pile buckles due to diminishing effective stress and shear strength owing to liquefaction, buckling instability can be a possible failure mechanism irrespective of the type of ground-level ground or sloping.

The mechanism and criteria to be used by practicing engineers are usually specified by prevailing codes. An example is bending failure assuming any non-liquefiable crust offers passive resistance and any liquefiable soil layer offers restraint equal to 30% of the overburden pressure (JRA 1996) as shown in Fig. 2. Furthermore, the Eurocode advises to design piles against bending due to inertia and kinematic forces arising from deformation of the surrounding soil (Bhattacharya 2003). It is required by the code to check separately against bending failure due to inertia load and not to add the effect of lateral spreading and inertia (Ishihara 1997). Unanimity among various researchers led to the assumption that lateral spreading is the cause of failure, such as Sato et al. (2001), Takahashi et al. (2002), Haigh (2002), Berrill (2001), Tokimatsu et al. (2001).

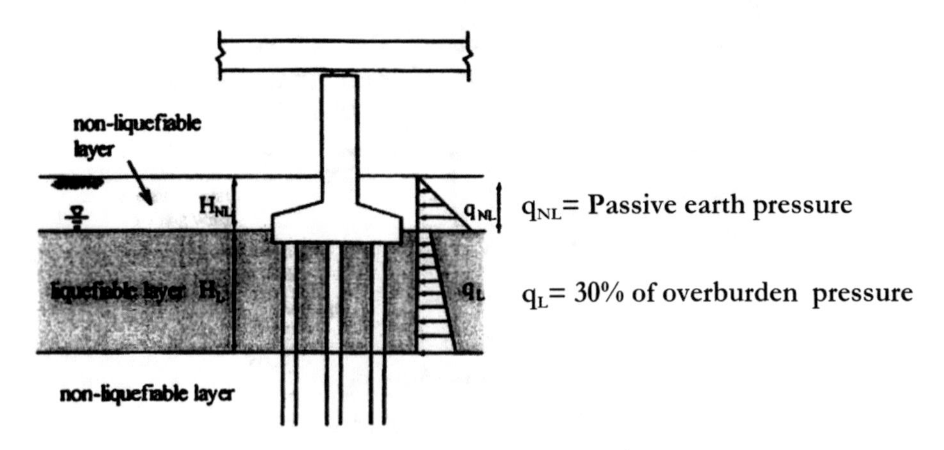

Fig. 2. The idealization for seismic design (JRA 1996)

2 Methodology

2.1 General Procedures for the Safety Assessment of Pile Foundations

Structures supported by piled foundations can be subjected to axial and lateral loadings. Structural failure of the pile can be seen in Fig. 3. Expected pile deformations under service loads and their ultimate load capacity are controlling factors in foundation design. There are several methods for calculating the bearing capacity of foundations including theoretical static analysis (Kulhawy 1984; Poulos 1989), procedures based on in situ test results (Meyerhof 1976; Fellenius 1997), dynamic methods (Rausche et al. 1985; Fellenius 2006), or interpretation of full scale pile load test (Fellenius 1990).

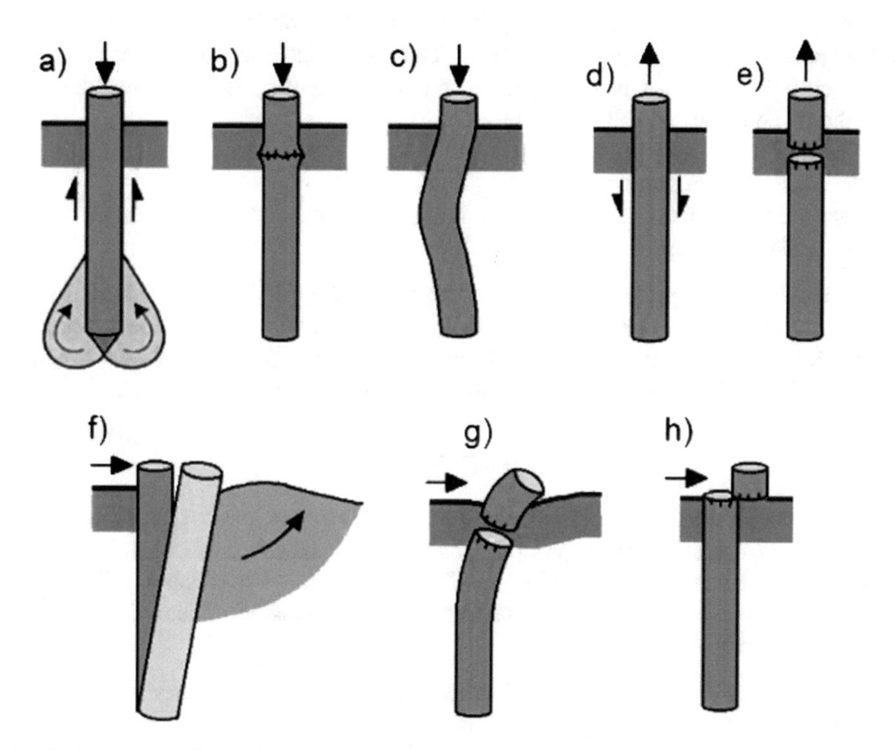

Fig. 3. Structural failure of the pile in different cases. (a)–(c) compression, (d), (e) tension, (f), (h) transverse (Wrana 2015)

2.1.1 Axial Bearing Capacity

The axial load capacity of the pile is derived from friction on the pile shaft and resistance at the pile tip. The ultimate bearing capacity Q_u of a pile installed into sand can be expressed as Eq. (1):

$$Q_{uc} = Q_p + Q_s + W_{ds} - W_p \tag{1}$$

where point resistance is Q_p, shaft resistance is Q_s, weight of the pile is W_p, weight of the displaced soil is W_{ds}. For calculating pile capacity in tension, Eq. (2) is used:

$$Q_{ut} = Q_s + W_p \tag{2}$$

where Q_{ut} is positive in tension and Q_s is positive downwards. The end bearing Q_p is taken zero in the case of tensile capacity. The pile weight, W_p in both of the above equations, should be net pile weight, i.e., the total weight of the pile minus the total weight of the displaced soil and water.

2.1.2 Lateral Resistance

Several methods are available for determining the ultimate lateral resistance to pile in cohesionless soil (Hansen 1961; Broms 1964):

(1) Hansen (1961) presented an expression for predicting the ultimate lateral resistance to piles in a general $c - \phi$' soil, where c and ϕ' are the cohesion and the effective internal friction angle of the soil, respectively. For a cohesionless soil, $c = 0$ and the ultimate lateral resistance can be calculated by Eq. (3):

$$p_u = K_q \gamma z B \tag{3}$$

where p_u is ultimate lateral resistance in the unit of force per pile length, K_q is Hansen earth pressure coefficient which is a function of ϕ', γ is the effective unit weight of soil, z is the depth from the ground surface and B is diameter or width of the pile. The lateral deflections have been computed assuming that the coefficient of subgrade reaction increases linearly with depth.

(2) Broms (1964) suggested the ultimate lateral resistance in cohesionless soil as following Eq. (4):

$$p_u = 3K_q \gamma z B \tag{4}$$

where $K_p = 45° + \phi'/2$ is passive earth pressure coefficient. Using Eq. (4), Broms (1964) prepared charts in a non-dimensional form giving the lateral capacity of piles in terms of the plastic moment and geometry of the pile.

2.2 General Procedure for the Safety Assessment of Piles in the Liquefied Ground

2.2.1 The Seismic Axial Capacity of Piled Foundations

The axial loading of piled foundations during earthquakes is complex, with the structure having to carry the vertical loads, which are applied under normal conditions, as well as additional axial load arising from the seismic excitation. A key feature of the end bearing capacity and shaft friction capacity was noted to be effective stress level in the soil profile, resulting in the loss of shaft friction and pile end bearing capacities as described by Knappett and Madabhushi (2008).

Bhattacharya (2006) indicated the static axial loading acting on each pile beneath the building is equally loaded during the static condition by neglecting any eccentricity of loading. During earthquake excitation, inertia action of the superstructure will impose dynamic loads on the piles, which can increase the total axial load on several piles, as given by Eq. (5):

$$P_{dynamic} = P_{static} + \Delta P = (1 + \alpha)P_{static} \tag{5}$$

Equation (5) needs the information of dynamic axial load factor α, which is a function of the type, dimension and mass of the superstructure, the characteristics of seismic shaking, as well as material properties and geometry of the pile foundation.

2.2.2 Lateral Resistance in Liquefied Ground

The lateral resistance of liquefying sand in the field undoubtedly depends upon numerous factors that are not yet fully understood or readily quantifiable. These factors can reasonably be expected to include everything that affects the stress-strain response of saturated sand, including relative density, drainage conditions, relative magnitudes of monotonic and cyclic loading components, number of loading cycles, and soil characteristics (Wilson 2000).

Takahashi et al. (2002) conducted an experimental work to study the lateral resistance of pile in liquefied soil. The test results showed that the initial resistance to movement is negligible at all rates of loading but some resistance was mobilized after some amount of displacement.

2.2.3 Pile Buckling as Soil Liquefies

Structure design with column buckling and beam bending criteria require different approaches. The former is based on strength and the latter is on stiffness. Bending failure depends on the bending strength, for instance moment at first yield (M_y), and plastic moment capacity (M_p) of the pile. Whereas buckling represents a sudden instability of the pile when axial load reaches the critical value (P_{cr}) described by Dash et al. (2010).

Bhattacharya (2015) indicated the static axial load at which a frame supported on slender columns becomes laterally unstable is commonly known as the elastic critical load of the frame of buckling load. The elastic critical load of the pile is defined by Eq. (6):

$$P_{cr} = \frac{\pi^2 EI}{L_{eff}^2} \tag{6}$$

where L_{eff} is the effective length of the column, which depends on the boundary condition at the column end i.e., fixed, pinned or free. The actual failure load $P_{failure}$ is therefore some factor ϕ ($\phi < 1$) times the theoretical Euler's buckling load given by Eq. (7):

$$P_{failure} = \phi P_{cr} \tag{7}$$

Based on Euler's theory, it may be inferred that buckling instability is initiated at around $\phi \doteqdot 0.35$.

As ground shaking starts, the excess pore pressure gradually increases which will in turn decrease the effective stress in the soil. As the effective stress of the soil approaches zero, the soil loses its strength and liquefies. Hence, the confining stress of the soil around the pile will be decreasing drastically, and eventually lead to buckling of the piles as shown in Fig. 4. Bending failure could be avoided by increasing the yield strength of the material. i.e. by using high-grade concrete or additional reinforcement, but it may not suffice to avoid buckling. To avoid buckling, there should be a minimum diameter depending on the depth of the liquefaction soil. In contrast, the pile has often being designed as a beam (Bhattacharya et al. 2004).

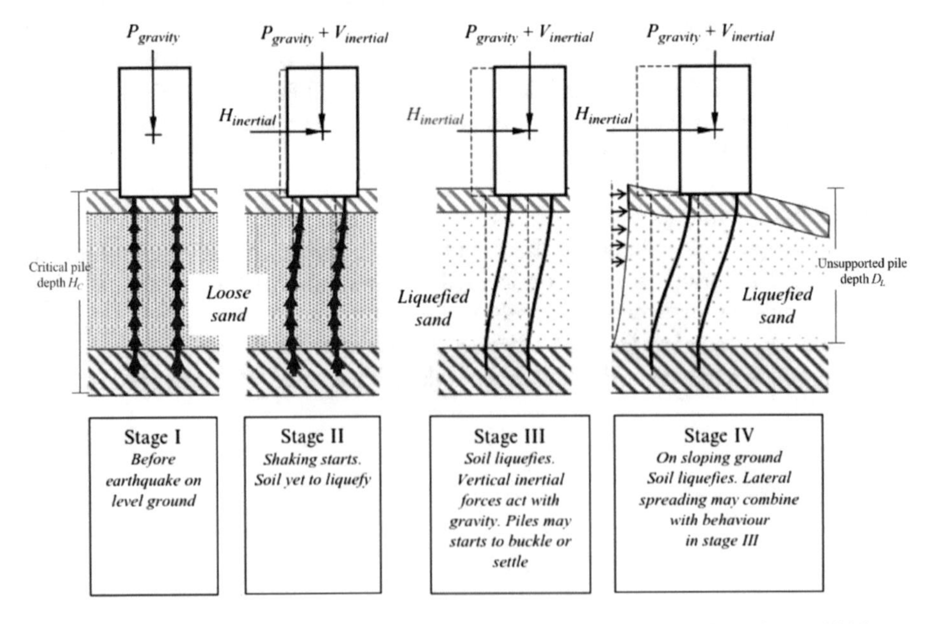

Fig. 4. Load applied to a pile foundation and failure mechanism (Bhattacharya 2015)

2.3 Analysis Procedures for Buckling Assessment of Piles Due to Soil Liquefaction

2.3.1 Liquefaction Assessment of Foundation Soil

Factors of safety against liquefaction could be computed using SPT-N based liquefaction evaluation procedure proposed by Youd et al. (2001). Liquefaction analysis calculates factors of safety against liquefaction at separated soil layer depths of a borehole. To address the severity of liquefaction for the entire borehole in the ground, the computed factors of safety and the associated depth intervals need to be considered. To evaluate the factor safety against liquefaction, both the soil's resistance to liquefaction and demand imposed on the soil by the earthquake need to be estimated. For the simplified approach used herein, the amplitude of cyclic loading is proportional to the peak ground acceleration (PGA) at the ground surface and the duration is related to the earthquake magnitude. The peak ground acceleration at SPT-N sites needs to be estimated for representative design earthquake.

The factor of safety against the initiation of liquefaction of a soil under a given seismic loading is generally defined as the ratio of cyclic resistance ratio (CRR), which is a measure of liquefaction resistance, over cyclic stress ratio (CSR), which is a representation of seismic loading that causes liquefaction. The term CSR is calculated in this paper as follows Eq. (8) (Youd et al. 2001):

$$CSR = 0.65\left(\frac{a_{max}}{g}\right)\left(\frac{\sigma_v}{\sigma_v'}\right)rd \tag{8}$$

where σ_v is the vertical total stress of the soil at the depth considered, σ_v' is the vertical effective stress, a_{max} is the peak horizontal ground surface acceleration, g is the acceleration of gravity, r_d is the depth-dependent shear stress reduction factor (dimensionless). The term CRR is calculated using SPT-N data. The following empirical equation developed by Youd et al. (2001):

$$CRR = \left(\frac{1}{34 - N_{1,60,FC}} + \frac{N_{1,60,FC}}{135} + \frac{50}{(10N_{1,60,FC} + 45)^2} - \frac{1}{200}\right)MSF \tag{9}$$

where $N_{1,60,FC}$ is the SPT blow count normalized to an overburden pressure of 1 atm, a hammer efficiency of 60%, and correction of fines content. MSF is the magnitude scaling factor for the adjustment of an earthquake magnitude of 7.5 to the magnitude of design earthquake of the site. The equation for factor of safety (F_L) against liquefaction is written as follows:

$$F_L = CRR/CSR \tag{10}$$

Iwasaki et al. (1982) developed the liquefaction potential index (LPI) to predict the potential of liquefaction to cause foundation damage at a site. The surface effect from liquefaction at depths greater than 20 m are rarely report, limited the computation of LPI to depth (z) ranging from 0 to 20 m. Proposed the following definition as Eq. (11):

$$LPI = \int_0^{20m} F.w(z)dz = \int_0^{20m} F.(10 - 0.5z)dz \tag{11}$$

In Eq. (11), $F = 1 - F_L$ for $F_L \leq 1$ and $F = 0$ for $F_L > 1$, where F_L is obtained from simplified liquefaction evaluation procedure. $w(z)$ is depth-weighting. Thus, it is assumed the severity of liquefaction manifestation is proportional to (1) the thickness of liquefied layer; (2) the amount by which F_L is less than 1.0; and (3) the proximity of the layer to the ground surface.

2.3.2 Unsupported Length Due to Liquefaction

The unsupported length of piles (D_L) indicated the extent along the pile where its lateral confining stress decrease significantly as a result of liquefaction of the surrounding soils. The unsupported length of piles is assessed based on the liquefaction profile. D_L is equal to the thickness of liquefied soil layers plus additional distance necessary for fixity into the upper or lower non-liquefied soil layer. The fixity is typically three to five times the diameter of the pile (Bhattacharya and Goda 2013). In this case study, we adopt a fixity of five diameters of the pile.

2.3.3 Buckling Assessment of Piles

The buckling assessment of piles is obtained based on the critical pile length (H_C). We assume that each pile is equally loaded during static condition, the static load (P_{static}) acts on each pile under the building. During shaking, the inertia of the superstructure imposes the dynamic axial load on the piles. Thus, the piles with increased axial load are perhaps vulnerable to buckle.

The critical pile length describes the minimum length that the pile will buckle due to axial load based on Euler's theory. As indicated above, the critical pile length is evaluated by considering the static and dynamic axial load as well as the boundary conditions of the pile as given at Eq. (5).

To determine of critical pile length, the limit state condition of failure is assumed, $P_{dynamic} = P_{failure}$ and $L_{eff} = H_c$, and Eq. (7) can be rewritten as:

$$P_{dynamic} = \phi P_{cr} = \frac{\phi \pi^2 EI}{K^2 H_C^2} \tag{12}$$

where EI is the bending stiffness of the pile and K is the effective column length factor depending on the boundary condition of the pile. In this study, ϕ value of 0.35 and K value of 1.0 are adopted in viewing that the pile head is fixed to the superstructure and the pile tip is embedded into the hard layer.

By rearranging Eq. (12), the critical pile length can be evaluated by Eq. (13):

$$H_C = \sqrt{\frac{0.35\pi^2 EI}{K^2 P_{dynamic}}} = \sqrt{\frac{0.35\pi^2 EI}{K^2(1+\alpha)P_{static}}} \tag{13}$$

Hence, liquefaction-induced pile buckling is indicated if $H_C < D_L$.

Finally, a buckling index G is adopted as the difference between the critical pile length (H_C) for buckling and the unsupported length (D_L) due to liquefaction of foundation soils, where H_C is the capacity variable and D_L is the demand term. Thus, the failure criterion can be indicated by Eq. (14):

$$G = H_C - D_L \tag{14}$$

As shown in Fig. 5, if G is greater than zero ($H_C > D_L$), then the pile is considered safe. Otherwise, the pile will be buckling due to seismic loading and soil liquefaction.

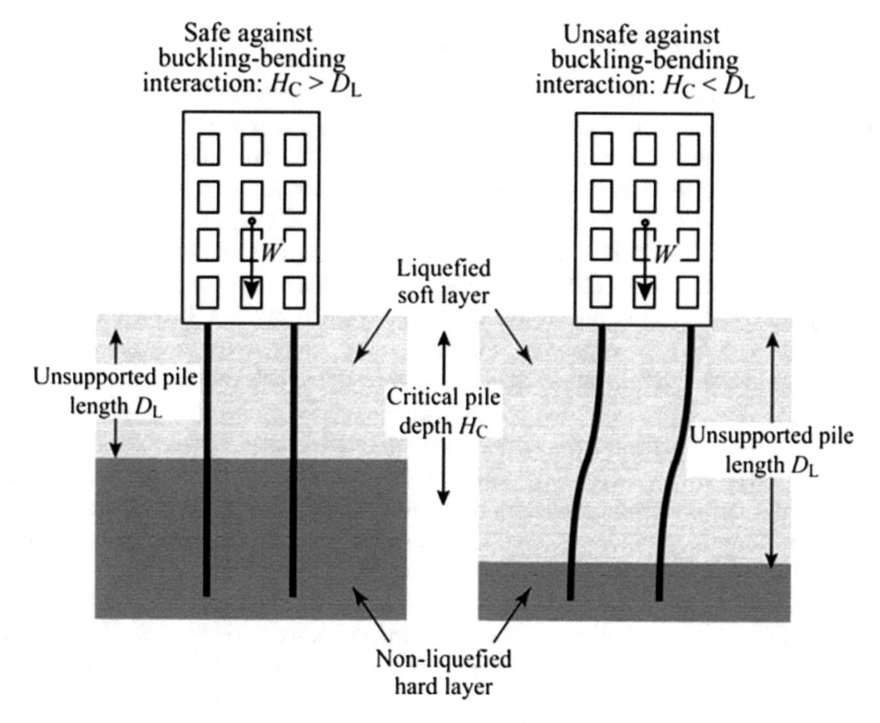

Fig. 5. Concept of critical length (H_C) and unsupported length (D_L), Bhattacharya (2015).

3 Case Study

3.1 Background Information

This case study related to a construction of a coal-fired power station in Central Java with capacity 2 × 1000 MW. In general, the project is located on an overlay alluvium deposit of the Muria mountain sediment material. The materials consist of coarse sand, fine sand, and clay. The soil contains old and recent river alluvial and shore deposits that were brought to the site by a small river flowing through and around the project site and tidal deposition. The upper soil deposits comprise alternating layers of very soft to soft clays and very loose to loose silty sands.

The case study covers 3 main facilities consisting of Boiler Units 5 & 6 and Central Control Building (CCB), which are supported by 1300, 1300, and 232 pre-stressed concrete piles, respectively. The piles are formed with exterior and interior diameters of 600 and 400 mm, respectively. The average pile length in Units 5 & 6 is 18 m, while in CCB is 12 m. The axial load for each pile (P_{static}) is 1450 kN, Young's modulus, E is 33.9 GPa, flexural rigidity, EI is 178 MN/m^2.

3.2 Liquefaction Analysis of the Ground

Several techniques on assessing liquefaction potential for the entire borehole depth have been proposed. The liquefaction potential of the site is evaluated based on the SPT-N approach by Youd et al. (2001) in association with the depth-weighted method by Iwasaki et al. (1982). As mentioned above, the earthquake with magnitude (M_w) of 6.8 and peak ground acceleration (PGA) of 0.210 g are adopted in the analysis by Youd's method. The PGA is adopted from the Indonesian spectra design-Puskim, Ministry of Public Works (2011) database and consistent with the regional ground motion. Based on limited on-site data, an energy ratio of 70% is assumed for the current study, which is consistent with the value adopted by sub-contractor soil investigation for the SPT hammer used. In an analysis, the unit weight of soil at each of the material strata is based on the borehole data obtained at the time of drilling.

In the project site, there were several monitoring wells used to know the fluctuation of groundwater levels. This study assumes the groundwater levels recorded in the borehole logs and monitoring wells for the analysis during SPT test and liquefaction. In order to determine groundwater levels the average groundwater data for the area is separated into dry and wet seasons. To account for seasonal fluctuations, 1.40 m and 0.90 m below the ground surface are assumed as the average groundwater levels for dry and wet seasons, respectively. An increase in the groundwater level would decrease the effective stress of soil, which would, in turn enhance the computed seismic force (i.e., CSR) at the depth of interest. On the other hand, as a result of an increase in the groundwater level, a decrease in the effective stress of soil would amplify the over-burden pressure correction factor in order to accommodate the underestimated SPT-N due to rising groundwater, and thus cause an increase in the cyclic resistance ratio (CRR).

Table 1. *LPI* category and numbers of borehole

LPI categories	Number of borehole computed	
	Wet season	Dry season
Low (0 < *LPI* ≤ 5)	6	7
High (5 < *LPI* ≤ 15)	4	5
Very high (15 < *LPI*)	6	4
Total	16	16

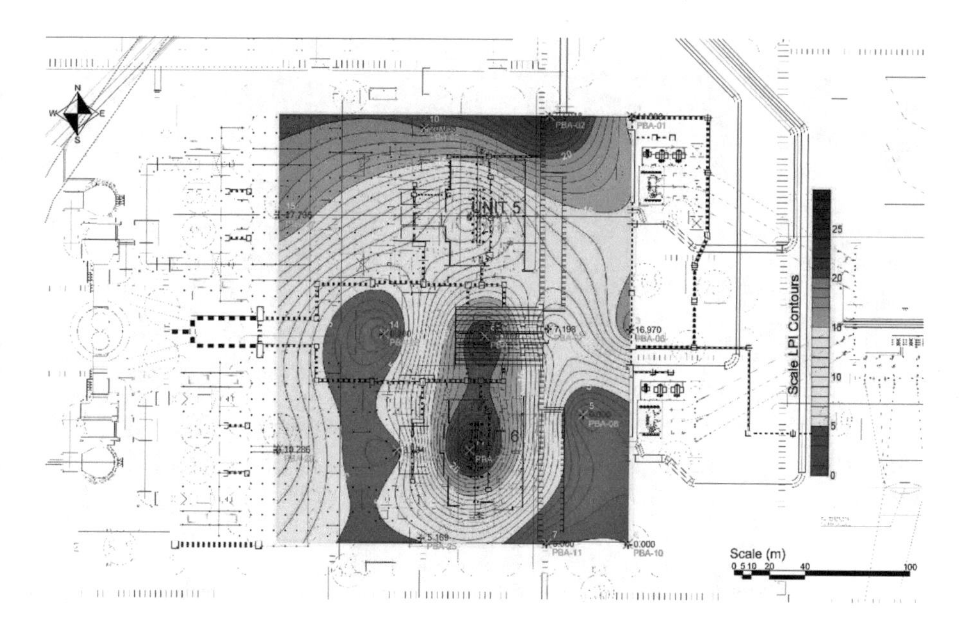

Fig. 6. *LPI* contours for groundwater scenario in wet season

Results of liquefaction potential and the unsupported pile length evaluation show the site is prone to liquefaction due to the design earthquake. Table 1 indicates the categories of liquefaction potential index (LPI) and the computed number of boreholes that cover the entire project site. Figure 6 shows LPI contour plots for the groundwater scenario during the wet season. The LPI results indicate more than 50% of the borehole with a LPI value greater than 5 (i.e., high to very high liquefaction potential) and the areas with high liquefaction potential generally fall in the center and southern parts of the site (CCB and Unit 6) and to north and east boundaries of Unit 5 (northern part of the site). The results of unsupported pile length from the representative 47 piles separated into 3 main facilities indicate more than 50% of the piles computed with D_L value less than 5 m. The areas with D_L value of more that 5 m generally fall in the center and southern part of the site, which are consistent with the LPI results with very high liquefaction potentials. These results are shown in Table 2 and Fig. 7.

Table 2. D_L category and numbers of pile

D_L categories (m)	Number of pile computed	
	Wet season	Dry season
$D_L \leq 5$	29	26
$5 < D_L \leq 10$	18	21
$10 < D_L \leq 15$	0	0
$15 < D_L$	0	0
Total	47	47

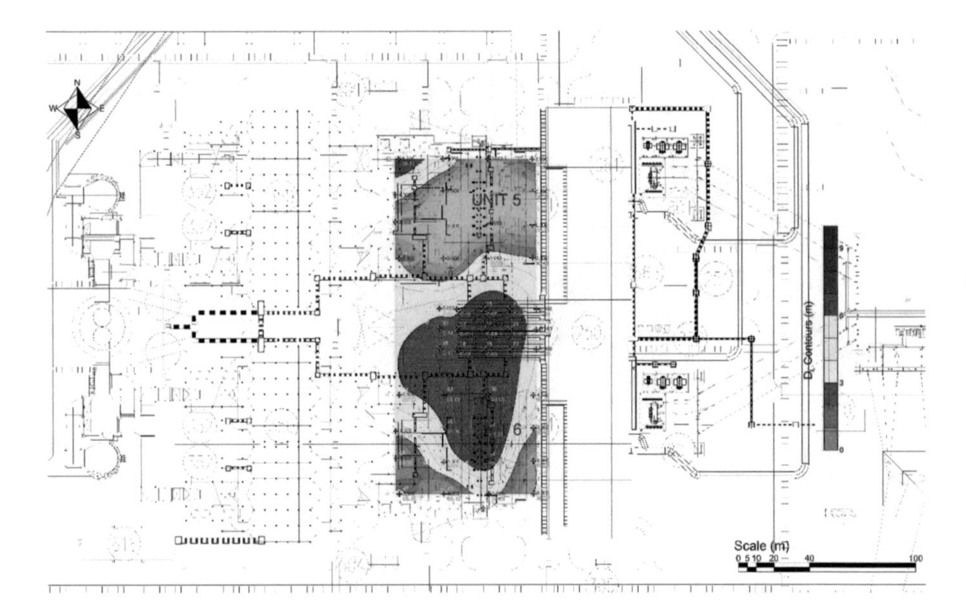

Fig. 7. D_L contours for post-liquefaction in wet season.

3.3 Buckling Analysis of Piles

The pile buckling analysis can be divided into two phases, pre-liquefaction and post-liquefaction. Since the pile will normally not buckle due to the confinement of foundation soils prior to liquefaction, the post-liquefaction phase is therefore considered in the buckling assessment of the piles in this study.

In analyzing the dynamic axial load of pile ($P_{dynamic}$) as indicated in Eq. (5) and to estimate the additional dynamic axial load (ΔP) on each of piles, the acting moment due to seismic shaking by the superstructure has to computed. Afterward, the moment can be distributed onto all of the resisting piles and the additional dynamic axial load on each of the piles can then be calculated.

Design provision BIS (2002) is the code of practice for estimating the force on the superstructure as well as the base shear, as given by:

$$V_B = C_s W \tag{15}$$

where W is the total dead load from superstructure and C_s is seismic response coefficient, for which C_s could be computed by the four parameters (1) Z is zone factor for maximum considered earthquake (MCE) (0.16; Prakash 2004; Bhatia et al. 1999), (2) I is importance factor for the structure (1.5 for coal-fired power plants), (3) R is response reduction factor, depending on the perceived seismic damage performance characterized by ductile or brittle deformations (4.0 for steel frame with concentric braces, (4) Sa/g is the average response acceleration coefficient, which is a function on the site and vibration period on the structure. The spectrum acceleration can be estimated based on the fundamental period of the structure. For the post-liquefaction situation, the fundamental period the structure can be calculated by:

$$T_{post} = 2\pi\sqrt{\frac{W/g}{N_p \times 12EI/D_E^3}} \tag{16}$$

where N_p is the total number of piles of the building, and is the depth to the lower boundary of liquefied soil plus and additional fixity. With the calculated fundamental period, the spectrum acceleration, and the base shear of the building as well, due to the design earthquake can then be obtained based on the design spectrum for the case of soft soil sites (BIS 2002).

In order to compute seismic moment on the superstructure, the arm where the inertial force acts can be estimated by the following:

$$ARM_{post} = D_E + \beta_3 H \tag{17}$$

where H is the height of the building, and β_3 is the coefficient to account for the effective height where the inertia acts in a post liquefaction condition (typically, 0.5). Henceforth, the base shear and the arm, the acting moment can be computed as:

$$M_{post} = V_B ARM_{post} \tag{18}$$

The overall moment is then distributed to all of resisting piles of the superstructure for computing the additional dynamic axial load on each of the piles. To do this, the utmost dynamic load for piles located at the peripheral boundary of the pile foundation needs to be calculated first, and then the additional dynamic load of the inner piles can be estimated by assuming the dynamic load is proportional to the distance between the pile of concern and the axis of symmetry of the foundation area.

By the assuming a rectangular arrangement of piles (n rows $\times$ m columns) and the acting moment in the direction of row, the additional dynamic axial loads of the peripheral and inner piles can thus be computed, respectively, as follows:

$$\Delta P_{max} = \frac{x_{max} M_{post}}{2n\left(\sum_{i=1}^{max} x_i^2\right)} \tag{19}$$

$$\Delta P_i = \left(\frac{x_i}{x_{max}}\right) \Delta \mathrm{P}_{max} \tag{20}$$

where x_{max} and x_i are the distances of the peripheral and inner piles, respectively, to the axis of symmetry of the foundation area. Finally, the dynamic axial load ($P_{dynamic}$) can then be calculated by Eq. (5).

With the evaluated dynamic axial loads, Eq. (13) can be applied to estimate the critical length for each of the piles. Table 3 and Fig. 8 show the results of H_C computations for the 47 representative piles that cover main facilities (Units 5 & 6 and CCB) of the site. The results indicate the minimum length that the on-site piles will buckle is 18 m, when the foundation soils are liquefied due to the design earthquake in wet season.

Table 3. H_C category and numbers of pile

H_C categories (m)	Number of pile computed	
	Wet season	Dry season
$H_C \leq 5$	0	0
$5 < H_C \leq 10$	0	0
$10 < H_C \leq 15$	0	0
$15 < H_C \leq 20$	34	34
$20 < H_C$	13	13
Total	47	47

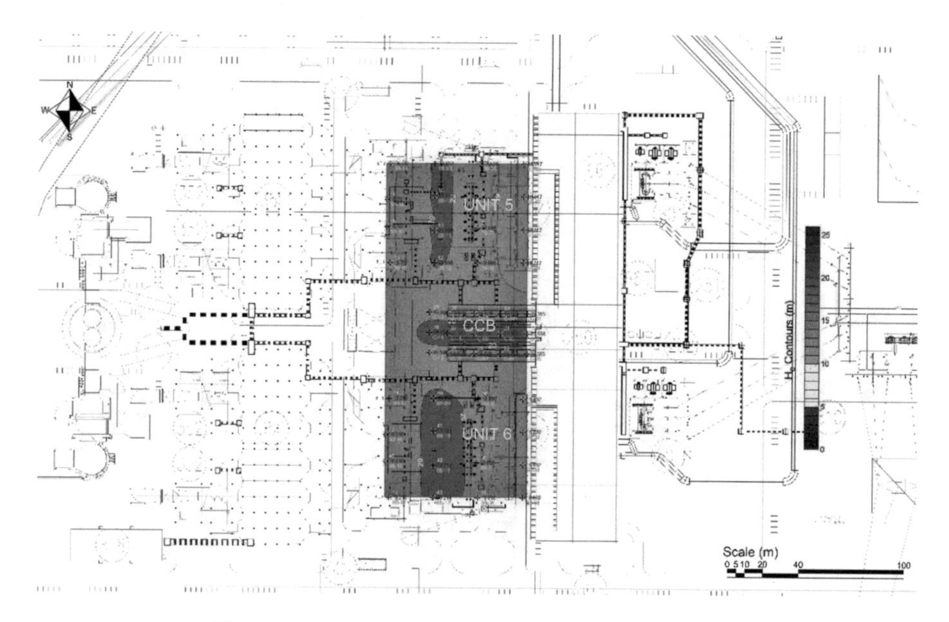

Fig. 8. H_C contours for post-liquefaction in wet season

3.4 Buckling Index Computation

Based on the above the analyses on the unsupported pile length (D_L) due to liquefaction and the critical pile length (H_C), the buckling index (G) can hence be assessed to determine if the piles are adequate in resisting the buckling instability per Eq. (14). As stated previously, if a G value is greater than zero then the pile is safe. Otherwise, the pile will be buckling. As shown in Table 4 and Fig. 9, the project site (Units 5 & 6 and CCB) will be safe from buckling instability, with calculated G values of greater than 10 (m) and an average of G value of 15 (m), for all of the piles at the project site.

Table 4. G category and numbers of pile

G categories (m)	Number of pile computed	
	Wet season	Dry season
$G \leq -5$	0	0
$-5 < G \leq 0$	0	0
$0 < G \leq 5$	0	0
$5 < G \leq 10$	0	0
$10 < G$	47	47
Total	47	47

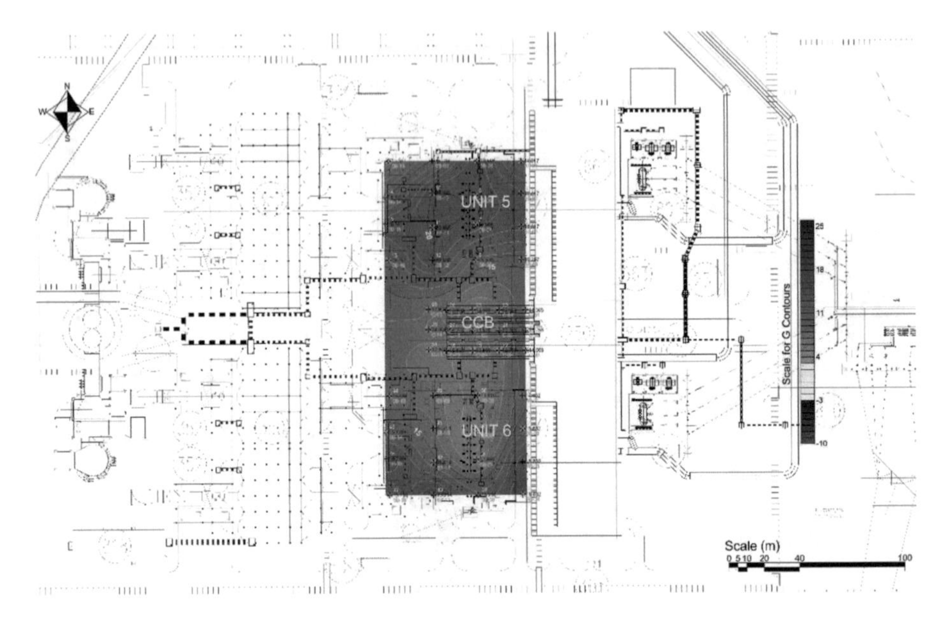

Fig. 9. *G* contours for post-liquefaction in wet season.

4 Conclusions

This study discusses the computation and assessment of liquefaction potential of foundation soils and buckling instability of piles for a coal-fired power station in Indonesia. Major findings of the study are listed as follows:

(1) Subsurface explorations reveal the soil deposit of the site generally consists of soft sandy silts or clays interbedded with loose fine sands to a depth of about 9 m. The deeper strata will be stiff and hard clayey soils.
(2) Liquefaction analysis indicates the foundation soils are prone to liquefaction due to design earthquake, with more than 50% of the boreholes assessed showing $LPI > 5$ (high to very high liquefaction potentials).
(3) Buckling assessment indicates the piles of the site will be safe from buckling failure, with computed critical pile lengths (H_C) of greater than 18 m and buckling indices (G) of greater than 10 (m), due to soil liquefaction by design earthquake, for the piles assessed at the project site.

Acknowledgments. The authors express their sincere thanks to Mr. Patrick W. Soule and Mr. Priya Purwanta for their contributions and review comments to this paper.

REFERENCES

Berrill, J.B., Christensen, S.A., Keenan, R.P., Okada, W., Pettinga, J.R.: Case studies of lateral spreading forces on a piled foundation. Geotechnique **51**(6), 501–517 (2001)

Bhatia, S.C., Kumar, M.R., Gupta, H.K.: A probabilistic seismic hazard map of India and adjoining regions. Annali di Geofisica, No. **42**, 1153–1166 (1999)

Bhattacharya, S.: Pile Instability during Earthquake Liquefaction. Ph.D. thesis, University of Cambridge, UK (2003)

Bhattacharya, S.: Safety assessment of existing piled foundations in liquefiable soil against buckling instability. ISET. J. Earthq. Technol. **43**, 133–147 (2006)

Bhattacharya, S.: Safety assessment of piled buildings in liquefiable soils: mathematical tools. Encycl. Earthq. Eng. (2015). https://doi.org/10.1007/978-3-642-361975-5_232-1

Bhattacharya, S., Goda, K.: Probabilistic buckling analysis of axially loaded piles in liquefaction soil. Soil Dyn. Earthq. Eng. **45**, 13–24 (2013)

Bhattacharya, S., Madabhushi, S.P.G., Bolton, M.D.: An alternative mechanism of pile failure in liquefiable deposits during earthquakes. Geotechnique **54**, 203–213 (2004)

BIS. Indian Standard, Criteria for Earthquake Resistant Design of Structures, Part I. General Provisions and Buildings, Bureau of Indian Standards, New Delhi, India (2002)

Broms, B.B.: Lateral resistance of piles in cohesive soils. J. Soil Mech. Found. Div., ASCE **90** (2), 27–64 (1964)

Dash, S.R., Bhattacharya, S., Blakeborough, A.: Bending-buckling interaction as a failure mechanism of piles in liquefiable soils. Soil Dyn. Earthq. Eng. **30**, 32–39 (2010)

Fellenius, B.H.: Guidelines for Static Pile Design: A Continuing Education Short Course Text. Deep Foundations Institute, Hawthorne (1990)

Fellenius, B.H.: Piles subjected to negative friction: a procedure for design. Geotech. Eng. **28**(2), 277–281 (1997)

Fellenius, B.H.: The red book-basics of foundation design (2006). http://www.fellenius.net/

Finn, W.D.L., Thavaraj, T.: Deep foundations in liquefiable soils: case histories, centrifuge tests, and methods of analysis. In: Proceedings of the 4th International Conference on Recent Advances in Earthquake Geotechnical Engineering, San Diego, California, 26–31 March 2001

Finn, W.D.L., Fujita, N.: Piles in liquefiable soils: seismic analysis and design issues. Soil Dyn. Earthq. Eng. **22**(9), 731–742 (2002)

GEER (2019). Geotechnical reconnaissance: The 28 September 2018 M7.5 Palu-Donggala, Indonesia Earthquake. Geotechnical Extreme Events Reconnaissance

Haigh, S.K.: Effects of Earthquake-Induced Liquefaction on Pile Foundations in Sloping Ground, PhD thesis, University of Cambridge, UK (2002)

Hamada, M., O'Rourke, T.D.: Case studies of liquefaction and lifeline performance during past earthquakes. Volume 1, Japanese Case Studies, Technical Report NCEER92-0001, State University of New York at Buffalo, Buffalo, USA (1992)

Hansen, J.B.: The ultimate resistance of rigid piles against transversal forces. Bulletin No. 12, Danish Geotechnical Institute, Copenhagen, Denmark, pp. 5–9 (1961)

Ishihara, K.: Geotechnical aspects of the 1995 Kobe Earthquake. In: Proceedings of the 14th International Conference on Soil Mechanics and Foundation Engineering, Hamburg, Germany, pp. 2047–2073 (1997)

Iwasaki, T., Arakawa, T., Tokida, K.: Simplified procedures for assessing soil liquefaction during earthquakes. In: Proceedings of Soil Dynamics and Earthquake Engineering Conference, pp. 925–939 (1982)

JRA: Design Specifications of Highway Bridges, Part V: Seismic Design. Japan Road Association, Tokyo (1996)
Kasbani. The geological agency emergency response team found the North Lombok fault, Central Jakarta. Volcanology and Geological Disaster Mitigation, Geology Agency, Ministry of Energy and Mineral Resources of the Republic of Indonesia (2018). (in Indonesian)
Knappett, J.A., Madabhushi, S.P.G.: Liquefaction-induce settlement of pile groups in liquefiable and laterally spreading soils. J. Geotech. Geoenviron. Eng. **134**(11), 1609–1618 (2008)
Kulhawy, F.H.: Limiting tip and side resistance: fact or fallacy. In: Meyer, R.J. (ed.) Proceedings of Symposium on Analysis and Design of Pile Foundations, San Francisco, American Society of Civil Engineers, New York, pp. 80–89 (1984)
Meyerhof, G.G.: Bearing capacity and settlement of pile foundations. J. Geotech. Eng. Div. ASCE **102**(3), 195–228 (1976)
Ministry of Public Works. Indonesian Design Spectra Application, Center Settlement Research & Development, Indonesia (2011). (in Indonesian)
Poulos, H.G.: Pile behavior-theory and application. Geotechnique **39**(3), 365–415 (1989). https://doi.org/10.1680/geot.1989.39.3.365
Prakash, V.: Whither performance-based engineering in India, ISET. J. Earthq. Technol. **41**(1), 201–222 (2004)
Rausche, F., Goble, G.G., Likins, G.: Dynamic determination of pile capacity. J. Geotech. Eng. **111**(3), 367–383 (1985)
Sato, M., Ogasawara, M., Tazoh, T.: Reproduction of lateral ground displacements and lateral-flow earth pressures acting on a pile foundations using centrifuge modelling. In: Proceedings of the 4th International Conference on Recent Advances in Geotechnical Earthquake Engineering and Soil Dynamics and Symposium in Honour of Professor W.D. Liam Finn, San Diego, California, 26–31 March 2001 (2001)
Takahashi, A., Kuwano, Y., Yano, A.: Lateral resistance of buried cylinder in liquefied sand. In: Proceedings of the International Conference on Physical Modelling in Geotechnics, ICPMG-02, St. John's, Newfoundland, Canada, 10–12 July (2002)
Tohari, A.: Characteristic of passive liquefaction in the Padang City based on the microtremor method. In: Proceedings of the Results Presentation of the Research at the Geological Research Center, Indonesian Institute of Sciences, LIPI. Bandung (2013). (in Indonesian)
Tokimatsu, K., Suzuki, H., Suzuki, Y.: Back-calculated p-y relation of liquefied soils from large shaking table tests. In: Proceedings of the 4th International Conference on Recent Advances in Geotechnical Earthquake Engineering and Soil Dynamics and Symposium in Honour of Professor W.D. Liam Finn, San Diego, California, 26–31 March 2001 (2001)
Unjianto, B.: The worst damage from the earthquake on Merapi Mountain sediment deposit. Suara Merdeka Cyber News (2006). (in Indonesian)
Wilson, D.W., Boulanger, R.W., Kutter, B.L.: Observed seismic lateral resistance of liquefiying sand. J. Geotech. Geoenviron. Eng. **126**(10), 898–906 (2000)
Wrana, B.: Pile load capacity-calculation methods. Studia Geotechnica et Mechanica, vol. 37, no. 4 (2015). https://doi.org/10.1515/sgem-2015-0048
Youd, T.L., et al.: Liquefaction resistance of soil: summary report from 1996 NCEER & 1998 NCEER/NSF workshops on evaluation of liquefaction resistance of soils. J. Geotech. Geoenviron. Eng., ASCE **127**(10), 817–833 (2001)

Strategy for Rehabilitation and Strengthening of Dam - A Case Study of Temghar Dam

Pravin Kolhe[1], Sunil Pradakshine[2], Gunjan Karande-Jadhav[3], and Anand Tapase[4](✉)

[1] Pune Irrigation Project Circle, Pune, India
pravinkolhe82@gmail.com
[2] Bhama Askhed Dam Division, Pune, India
sunil.v.pradakshine@gmail.ccm
[3] Accelerated Irrigation Benefit Program Special Cell, Pune, India
gunjankarande@gmail.com
[4] Rayat Shikshan Sanstha's, Karmaveer Bhaurao Patil Collage of Engineering, Satara, India
tapaseanand@gmail.com

Abstract. Construction of Temghar Dam took place in stages from 1997 to 2010 due to the number of difficulties including forest clearance and allied issues. Construction work was completely stopped in July 2002 after creating a storage capacity of around 63%, which was regained back in the year 2009 almost after seven years and completed in the year 2010. Leakage was observed first time during its construction stage itself in the year 2001 which was observed increasing day by day as the reservoir water level increases. In the year 2016-17 a considerable leakage of around 2587 lps was noticed, so to deal with the leakage problem and to ensure the safety of dam it was proposed to provide the remedial repair measures in two stages i.e. long term and short term. Short term remedies includes, U/s drilling & grouting from top to inspection gallery and Inspection gallery to foundation gallery, drilling & grouting on D/s surface, cleaning vertical porous drains, curtain grouting in remaining portion in foundation gallery, monolith joint treatment, and U/S treatment – shotcrete with mesh and anchor bars into the masonry. Long term remedies include concrete backing cement grout consisting of fly ash, silica flume and admixtures such as SPG 1A, ACC2025 and ADMGP1 is used for grouting after studying and testing other grout options. Apart from laboratory tests, Marsh Cone Test, settlement test, pH, compressive strength and density test were carried out at the site. Various type of machinery such as Drilling machine Water intake test machine with single drum, grouting machine with double drum mixer & with RPM 1500 to 2000, Grout pump, water meter, pressure meter and batching plant with SCADA system are used in grouting process. Leakages before and after grouting were observed in terms of Lagoon values. Observations are recorded at different R.L.s and at different chainages to get an exact picture of leakages. It was observed that almost 80% leakage was stopped by June 2018. With the help of correct implementation of grouting techniques, leakages through masonry dam can be controlled however grouting or any other remedial measure is not an alternative to poor workmanship and quality control.

Keywords: Leakage · Masonry dam

H. El-Naggar et al. (Eds.): GeoMEast 2019, SUCI, pp. 107–120, 2020.
https://doi.org/10.1007/978-3-030-34252-4_9

1 Introduction

Introduction: Temghar Dam, salient Features, Construction period, Current status, etc. Temghar dam has been constructed across Mutha River near Pune district of Maharashtra during the period March 1997- May 2010. The Temghar project mainly caters to Domestic water supply to Pune city and irrigation of 1000 Ha agriculture land through K.T. weirs. Hydropower is also contemplated at the foot of the dam. The total utilization of 3.708 T.M.C. is planned for this project which is accommodated within 599 T.M.C. of Krishna water allocated to Maharashtra.

Temghar dam is a stone masonry dam with 5 m thick colgrout masonry (1:3) septum on u/s face, at bottom of thickness 3 m and on the downstream side as a triangular toe from the foundation up to RL 667 m where the stresses were more than 120 T/m^2. The total length of the dam is 1075 m and the maximum height is 86.6 m. It comprises 72 m long spillway portion in the centre from RD 528 to 600 m and non-overflow portion on either flank. The dam top RL is 711.40 m and Full Reservoir Level is at RL 706.50 m. Gross capacity at FRL is 108 Mcum i.e. 3.71 TMC. Silent features of the project are introduced in Table 1.

Table 1. Salient Feature

Sr.No	Particulars	Details
1.	source/Name of River	Mutha River
2.	Location:	
	State	Maharashtra
	Region	Western Maharashtra
	District	Pune
	Taluka	Mulshi
	Village	Temghar
	Toposheet No.	47 F/7, 47 F/11
	Latitude	18° 27' 0" (N)
	Longitude	73° 32' 0" (E)
3	Controlling Levels	In Meters
	River Bed Level	641.56
	M.D.D.L.	661.40
	F.S.L.	706.50
	H.F.L.	710.12
	T.B.L.	711.40
4.	Command Area(on d/s of the dam to backwater of Khadakwasla Project)	
	1. Gross command area	2000 Ha./4940 Acres
	2. Culturablecommand area	1600 Ha./3952 Acres
	3. Irrigable command area	1000 Ha./2470 Acres

1.1 Issues: Delay in Construction, Artificial Sand, Poor Quality, Leakage, Density, Etc

The construction of Temghar Dam, Pune India was started in March 1997. As per tender schedule, construction period was 44 months and work was expected to be completed by November 2000. However, an extension for completion was granted due to the increase in quantities and cropping up of extra items during execution. The construction of the dam was going in full swing from March 1997 to December 2001. However, work was totally stopped by the forest department in January 2002 due to the 4.5 ha of forest land coming under submergence. After receiving the due permission from the forest department, construction was again resumed in April 2009 and the dam was completed in 2010.

The upstream septum of 5 m width is provided of colgrout masonry to prevent leakage. Also on downstream colgrout masonry zone is provided at toe portion up to RL 667 m. Leakages in Temghar dam were observed at the time of construction which increased with an increase in reservoir storage level. Table 2 shows how leakage increase year by year.

Table 2. Yearly observations of leakage

Sr. No	Year of Construction	Max. total seepage in lps	Remarks
1	June 2000	–	
2	2001-02	72	
3	2002-03	58	
4	2003-04	47	
5	2004-05	68	
6	2005-06	90	
7	2006-07	395	
8	2007-08	430	
9	2008-09	524	
10	2009-10	508	Construction up to FRL
11	2010-11	526	
12	2011-12	602	Grouting is done in 2012
13	2012-13	307	Leakages observed after Grouting is done in 2012
14	2013-14	402	Leakages observed in the monsoon season of 2013
15	2014-15	385	
16	2015-16	1174	
17	2016-17	2581	
18	2017-18	1039	Leakage reduced from 2587 to 1039 lps after grouting of 2000 M.T. grout
19	2018-19	413.80	Leakage reduced from 2587 lps to 413.80 lps around 80% after grouting of 18000 M.T.

From the leakages and visual inspection, it is evident that particularly left flank portion and half of the central gorge portion work has been executed very badly. Upstream surface undulations and uneven slopes speak of bad workmanship. Cement slurry accumulated on the upstream surface, percolated from the shuttering for colgrout masonry reveals that cement grout is not able to occupy the space between the stones due to improper packing between the stages of masonry. Large cavities show the improper laying of stones, single drum mixer was used instead of double drum mixer which is not able to create a colloidal state of the mix. Also, the use of improper grading of artificial sand used seems to be responsible for forming inhomogeneous masonry. Crushed sand was used for the construction of colgrout masonry. Due to angular surfaces in the crushed sand together with high fineness modulus of the sand might have resulted in a non-colloidal state of colgrout slurry during construction and may have caused segregation of sand and cement. This may be one of the reasons for the inadequate strength of colgrout mortar and high seepage. Figure 1 shows the severity of leakages through the dam body.

Fig. 1. Photos showing leakages in Temghar Dam

1.2 Potential Hazards

Temghar dam is located on upstream of Khadakwasla dam, Pune India, from where drinking water is supplied to Pune city by a pipeline and lake water is also released into Khadakwasla RB canal to irrigate drought-prone area in the East. Water stored in Temghar dam is planned to be released back into the river for its use from Khadakwasla dam. Since Temghar dam is located on u/s of Khadakwasla dam, which was located on u/s of Pune city, it had high hazard potential in case of any mishap. Leakages from masonry induce leaching of free lime in cement rendering loss of strength of masonry and also loss of water and creating fear amongst the people residing on downstream.

1.3 Rehabilitation and Strengthening Strategy

To suggest the remedial measures for this leakages Government of Maharashtra decided to constitute an Expert committee for study and recommendations. So Temghar Dam Expert Committee (TDEC) was constituted vide Executive Director, Maharashtra Krishna Valley Development Corporation, Pune's order no. 1901 of 2014 issued under no. 6549, date 27.8.2014 to examine causes of seepage through Temghar Dam and exploring possible measures to reduce it and to access structural stability of dam as well as to suggest the remedial measures for its rehabilitation.

1.4 Findings of Ranade Committee

Temghar Dam Expert Committee (TDEC) studied the behavior of the Temghar dam by carrying out field inspection and reviewed record related to dam construction & quality control aspects. TDEC held discussions with the present site officials to know developments in seepage from the dam and seepage control measures tried by them. It was observed that most of the seepage was coming out in the form of jets at many places from dam body on the LB. In that comparison, seepage from gorge and RB was much less. Seepage was also coming from some monolith joints and foundation drains. TDEC felt need to improve the crude unscientific methods of measurement of seepage followed by the field officers at the site. It was felt necessary to find out the percentage of dissolved salts from seepage water & lake water, to know the loss of dissolved solids from dam masonry. Appropriate suggestions were given to the project staff in that matter. Study of 'Tomography' was carried out for the assessment of the presence of saturated zones in the dam masonry on LB. Some in-situ masonry samples also were tested for its density and crushing strength and finally, TDEC gave their opinion about remedial measures to reduce seepage in two broad categories wise Short term measures and long term measures.

A. **Short term remedial measures**

U/s surface treatment, dam body grouting, foundation curtain grouting, cleaning of porous block, repairing monolith joint.

B. **Long term measures**

Strengthening by post-tensioning cables, strengthening by earth backing, providing additional drainage gallery, masonry or concrete backing, buttressing.

1.5 Repair Work-Important Items

Short term measures such as grouting, providing effective workable treatment to the upstream face of the dam, completing remaining curtain grouting and dam body grouting, Rimming/drilling of porous drains to improve their functioning, etc. are needed to be addressed immediately.

Considering the severity of the problem and hazard potential of the dam it is planned to adopt immediate short term measure before coming monsoon to plug excessive leakages which may lead to the potential danger of piping and Long term measures are planned to provide a complete solution for leakages as well as strengthening of the dam.

1.6 Grouting

a. Grouting Pattern-
It was decided to go for exhaustive grouting. So drilling and grouting work was immediately started from dam top to Inspection Gallery, Inspection gallery to foundation gallery, from D/s as well from U/s side from April 2017.

Grouting from upstream face: Grouting the upstream colgrout septum from foundation to dam top is done by drilling 38 mm to 51 mm dia holes having depth at 3 m c/c both ways. Intermediate holes are drilled for water intake test and for grouting wherever necessary.

Grouting from Downstream face: Grouting of UCR masonry is done by drilling 51 mm to 75 mm dia holes with lightweight hydraulic diamond drilling machines (by percussion drilling). These are mounted on continuous support systems anchored firmly on dam top with anchor fasteners for maximum depth to cover all the masonry perpendicular to the downstream slopping face of the dam. Primary grouting at 6 m c/c and intermediate secondary grouting at 3 m c/c and tertiary at 1.50 m c/c is adopted.

Dam body grouting is also done from dam top to inspection gallery and from inspection gallery to foundation gallery wherever it was possible by drilling slant holes. Directional grouting it also carried out targeting maximum penetration under low pressure with reservoir empty condition. Stage grouting method is adopted. Grout pressure is given between 1.5 to 3.0 kg/cm^2. Location and schematic grouting pattern are illustrated in Table 3 and Fig. 2.

Table 3. Grouting pattern (Location of grout holes)

Sr. No.	Location	Pattern	
1	Dam Top	Ist Row, 1.5 m from U/S	IInd Row, 3 m from U/S
2	Inspection Gallery	Ist Row, 0.4 m from U/S	IInd Row, 0.1 m from D/S
3	Upstream side	(Primary, Secondary, Tertiary)	
4	Downstream side	(Primary, Secondary, Tertiary)	

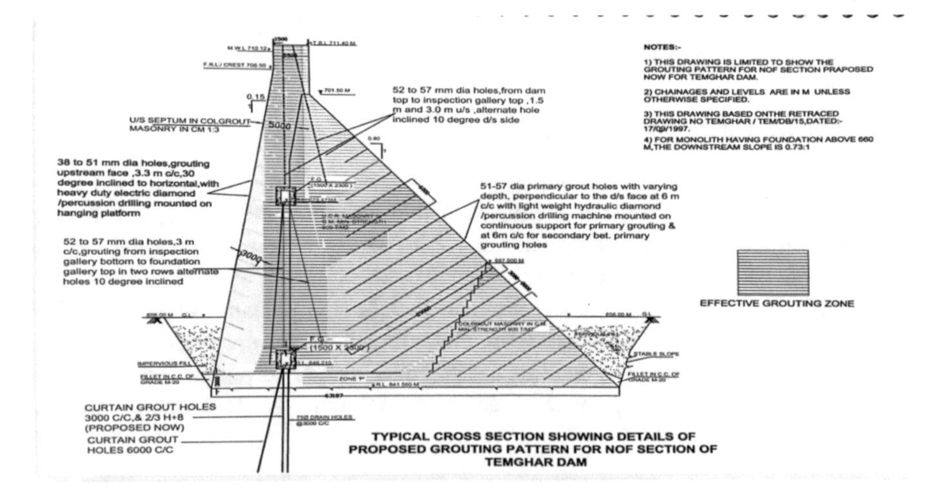

Fig. 2. Grouting pattern

b. Grout Mix Design-

A specially designed grout given by CWPRS, Pune after study of many other options, is used for grouting. Mix design for the same is shown in Table 4.

Table 4. Grout mix design

Cement:fly ash: sillica	W/C ratio	Admixture % by the weight of cementitious material			Avg. compressive strength (Kg/cm^2)		Density gm/cc	Marsh cone time sec	Settlement settlement
		Powder	Superplasticizers	Accelerators	7 days	28 days			
80:18:2	0.65	2	5 ml per kg of cementatious material		107	222	2.8	27 to 36	Less than 5%

1.7 Significance of Each Ingredient in the Grout Mix

Fly Ash - It improves fluidity reduces the heat of hydration of grout.

Silica Fume – It enhances the density and strength of grout.

Admixtures

SPG 1A- It enhances flow & penetration characteristics of the grout with a higher solids loading.

ACC2025 - It accelerates the setting of cement & facilitates speedier operation.

ADMGP1- It enhances bonding & sealing properties and underwater sealing ability against washout effects.

Various grout mixes were tested in CWPRS laboratory and cores are also extracted for testing purposes. Testing at CWPRS Pune is shown in Figs. 4 and 5.

Fig. 4. Masonry blocks casted for trail purpose @ CWPRS

Fig. 5. Cores extracted from test blocks for testing @ CWPRS

c. Important machinery used for grouting work at the actual site-
Drilling machine Water intake test machine with a single drum Grouting machine with double drum mixer & with RPM 1500 to 2000 (Mixing with high-shear mixers is extremely efficient and can be completed rapidly ($\sim$1 min) after all components have been added to the mixer. The high shear imparted to the grout generates substantial heat, which accelerates set time. Therefore, mixing time should not be excessively long and must be controlled) Grout pump Water meter, pressure meter, etc. Winch, hanging platform Compressors Water pump Various types of equipment used for grouting are operations are shown in Figs. 6, 7, 8 and 9. Work of grouting started in April 2017. Before that leakage observed was 2587 lps and now in 2018 leakage observed are 413.80 lps. Monolith wise quantity of grouting executed up to 2018 is as shown in Table 5.

Table 5. Monolith wise quantity of grouting executed

Monolith No.	Executed quantity of grouting (MT)
1 to 13	10490.38
14 to 17	6367.14
18 to 27	979.00
Total	17836.52

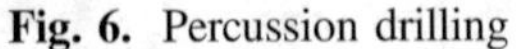

Fig. 6. Percussion drilling

Fig. 7. Core barrel drilling

Till date, about 50% of the work is completed and with this work, about 80% of the leakages are reduced. As per the Temghar Dam Expert Committee (TDEC) out of five short term recommendations only grouting and that is also in partial quantities are

Fig. 8. Grouting machine

Fig. 9. Checking RPM of grouting machine

executed. But with this, about 80% of leakage has come down. After the execution of all grouting work as well as remaining measures like upstream treatment, Vertical porous hole cleaning, Curtain grouting, repairing monolith joint, etc. it can be said that leakage will be reduced within the permissible limits.

d. Tests and quality control

To maintain the quality of work there are various tastes conducted on the grout, which includes marsh cone test, settlement, pH value, strength, and density.

Significance of various tests

Marsh cone test- To check the flow time of grout marsh cone test is carried out and the permissible flow time should be between 28 to 30 s.

Settlement test- For assessing the colloidal stage of slurry settlement test is taken. The settlement should be below 5%.

pH test- To check the alkalinity of the grout & in turn, the chemical composition of the ingredient material pH value is determined. Permissible pH value of grout should be between 11 to 13.

Compressive strength- To assess the strength of the masonry compressive strength is measured. It should be more than or equal to 222 kg/cm^2.

Density – This test is important from a safety point of view. It should be more than or equal to 1.8 gm/cc.

Standardize formats are maintained at the site in which periodical reports (weekly, fortnightly, monthly and seasonal) about physical progress, cement consumption, mixer registers, field tests, laboratory tests, etc. are recorded by engineers at different levels.

Till now, about 50% of the work is completed and the leakages are curtailed to the tune of 80%. It is planned to complete whole work in coming 1 & ½ years & by the end of May-2020 whole work along with all measures is expected to be completed. It is also planned to check the density of post grouting masonry by nuclear borehole logging and also test the weak zones by Tomography. After these results, it is planned to go for permanent measures like concrete backing of the dam as suggested by Temghar Dam Expert Committee. Statement of various field tests carried at the site is shown in Table 6.

Table 6. Statement of field tests carried out at the site

Sr. No.	Test	Results		Frequency	
		As per CWPRS guidance	As per observations at site	As per EM 1110-2-3506	As per actual at site
1	Marsh Cone Test	27 to 36 s	28 to 35	Once per day	One per every 2 h per mixer
2	Settlement Test	<5%	2 to 3%	Once per week	One per every 2 h per mixer
3	pH	11 to 13	11 to 13	–	One per every 2 h per mixer
4	Compressive Strength	222 kg/cm^2	More than 250 kg/cm^2	At the time of mix design	One per day per mixer
5	Density	1.8 gm/cc	More than 1.8 gm/cc	–	One per day per mixer

1.8 Comparison of Leakage After Grouting

Leakages are measured at various stages during grouting. Table 7 shows a comparison of leakages in various years and Fig. 10 shows graphical representation for the same. Figure 11, 12, 13, 14, 15, and 16 shows actual site photographs showing a comparison of leakage before and after grouting.

Table 7. Leakage comparison after grouting

Sr. No.	Year of Construction	Max. total seepage in lps	Remarks
1	2016-17	2587	After fixing & measuring the seepage by Trapezoidal notches on d/s side
2	2017-18	1039	Leakage reduced from 2587 to 1039 lps at the same level
3	2018-19	413.80	Leakage reduced from 2587 lps to 413.80 lps at same level Which is around 84%

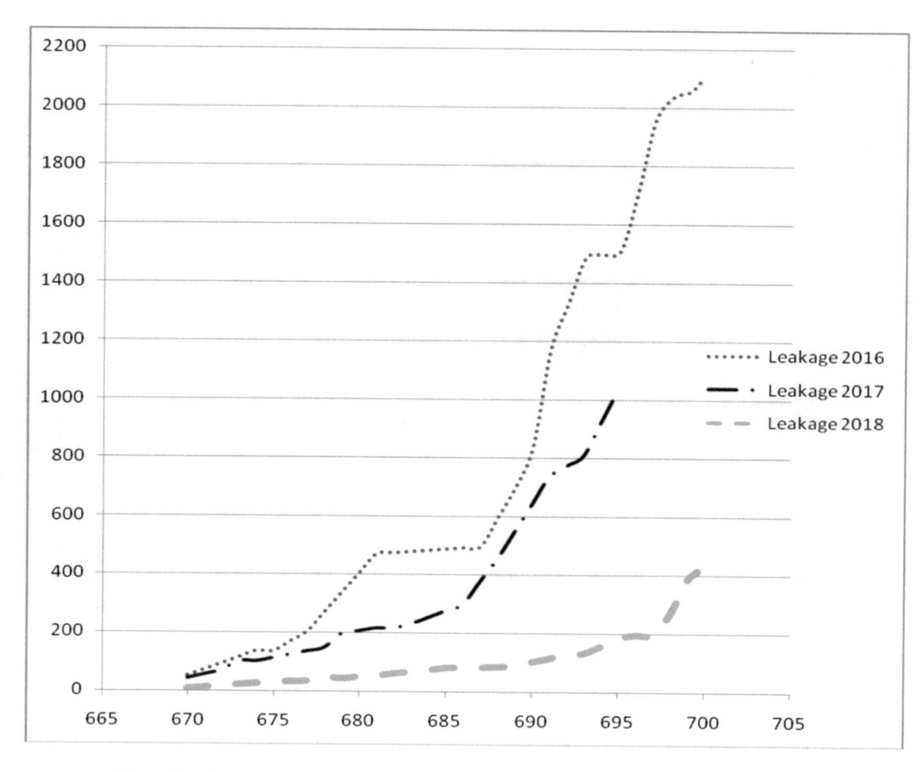

Fig. 10. Graph showing leakage comparison at reservoir filling condition

Fig. 11. D/S side of M-10, M11, M12 before grouting

Fig. 12. D/S side of M-10, M11, M12 after grouting

Fig. 13. D/S side of M14, 14A before grouting

Fig. 14. D/S side of M14, 14A after grouting

1.9 Shotcrete Treatment to U/S Face

Temghar Dam Expert Committee suggested treatment of upstream face as one of the short term measures in the repair of Temghar Dam. The committee also suggested a sequence of activities to stop leakage from the dam which is as follows.

Fig. 15. D/S side of M14, 14A before grouting

Fig. 16. D/S side of M14, 14A after grouting

Grouting- Dam top to inspection gallery Grouting- Inspection gallery to foundation gallery, Grouting- Upstream face, Treatment of upstream face to make it waterproof Curtain grouting in the month of April and May when the water level is reservoir is low.

In order to decide the method of upstream face treatment, Expert committee suggested applying 10 mm thick layer of UV resistant non-shrink cementations repair mortar on the upstream surface after thoroughly cleaning dam surface and removing all loose materials from the surface of the dam. Committee further suggested applying seal coat over this layer. In order to have proper bonding of UV resistant non-shrink cementations repair mortar with parent dam body bond coat is also suggested by the expert committee.

As per suggestions of the committee in the year 2015 the above treatment is done on left side flank of the dam in the patches of 2 m × 2 m at Chainage 415 to 466 at RL 665 to 669. For this treatment Poly Ironite Ceramic Cementitious (PICC) is used as a base coat and Poly Cementitious material is used as a seal coat. Observations of the above patches are taken from time to time. However, it was observed that the PICC layer is detached from the parent dam surface and cracks were also seen on it. At some places, scaling is also observed. After these observations, it was decided that by the committee to discard this treatment for repair and instead of it, it is decided to use poly fiber reinforced shotcrete on the upstream face. As per the various experiments carried out at the field with the help of CWPRS, it is decided to give polypropylene fibre reinforcement shotcrete treatment to U/s face of the Dam. For this purpose following provisions are made for wet shotcrete treatment to U/S surface anchor bars of 25 mm dia. 1.5 m length anchored at 1 m × 1 m distance.

To give the strength to shotcrete layer wire mesh of 13 gauge of 50 mm × 50 mm is to be used. To create a bond between bars and wire mesh, an m.s. square plate of 100 mm × 100 mm size and 5 mm thickness is being fixed on outer end by threading. Mix design provided by CWPRS Pune is used for shortcret which is given in Table 8.

Table 8. Mix design for shortcrete

Sr. No.	Ingredient	Weight in Kg
1	P.P.C. cement	376
2	Fly ash	70.5
3	Micro silica	23.5
4	Crush sand passing through 4.75 mm IS sieve	1254
5	Aggregates (particle size 4.75 mm to 10 mm)	626
6	Polypropylene fiber	1.17
7	Accelerator	4.70 (1% by weight of Cementitious material)
8	Plasticizer (SPG- super plasticizer)	4.70 (1% by weight of Cementitious material)
9	Water	329 L

2 Conclusions

With the correct implementation of short term measures suggested by Temghar Dam expert committee, which mainly consist of dam body grouting and foundation curtain grouting, about 80% leakages from dam body were stopped till June 2018.

Before implementation of any grouting procedure, proper mix design is mandatory with due consideration of actual field requirements. Mix design developed in the laboratory must be actually tested on actual site before using it on large scale.

Grouting pattern also plays vital role in successes of grouting. Grouting pattern must be carefully planned so that whole area is covered with grout without leaving any void. Proper quality control at field as well as in laboratory is an essential requirement for effective grouting and its desired results. For carrying out shortcrete operations especially on inclined faces (such as upstream face of dam) various options must be tested at actual site to choose best possible and economic option to suit particular site conditions. Record of each and every test, findings etc. must be kept for further comparison and to improve quality of work. With successful implementation of grouting techniques, leakages through masonary dam can be controlled.

References

1. Mix design & correspondence with CWPRS
2. I.S. 6066 1994 Pressure grouting of the rock foundation
3. I.S. 11293 1993 Design of grout curtains
4. I.S. 11216 1985 Permeability test for masonry
5. I.S. 3812 2013 Specifications for pulverized fuel ash
6. I.S. 15388 2003 Specifications for Silica Fume
7. I.S. 1727 1967 Methods for test for pozzolanic material
8. ASTM D 6910 – 04 Standard test method for Marsh Funnel Viscosity
9. US Army code EM 1110-2-3506

Factors Affecting Lubrication of Pipejacking in Soft Alluvial Deposits

Wen-Chieh Cheng(✉) and Ge Li

Xi'an University of Architecture and Technology, Xi'an 710055, China
{w-c.cheng, lige}@xauat.edu.cn

Abstract. Inadequate lubrication during pipe ramming can result in exaggerated jacking force leading to damages to the jacked pipe string and adjoining properties. However, a consensus due to limited studies has not yet reached. This study describes a method that can be used to evaluate the lubrication performance by the reduction in the frictional coefficient μ as a function of injection mode, soil and lubrication natures, and pipe deviation. The results of an application of the method to a pipe ramming project in soft alluvial deposits are presented. Compared to 0.4 kPa of the 2–8 m section of the gravel, the excessive pipe deviation, while spanning through the 8-21 m section of the same gravel at Drive C, increases the frictional stress τ_{ld} to 12.5 kPa, thereby reducing the reduction in the frictional coefficient μ to 71%. At Drive D, the combined effects, resulting from the excessive pipe deviation and the varying face resistance induced by driving into the 11–24 m section of clayey gravel, contribute to the growing frictional stress τ_{ld} of 4.0 kPa and are deemed as being the major cause leading to the low reduction of 84%. While pipejacking through the 24–31 m section of clayey gravel, the occasional gravel leads to the inability of developing lower face resistance causing the misleading reduction of 60%. To summarise, the excessive pipe deviation and/or the varying face resistance largely affects the percentage reduction in the μ value despite the excessive volumes of injected lubricant. The occasional gravel leads to the misleading reduction.

1 Introduction

The pipejacking technologies are specially favourable for urban pipeline systems construction because of their low cost and high efficiency. Despite some grouting technologies available for property protection purposes, the structural behaviour of constructed pipe string and joint between pipes require to be studied to enhance its resilient ability as subjected to differential ground movements resulting from adjoining deep excavations. Also, excessive jacking force or inadequate jacking capacity is often seen during pipejacking works leading to damages to adjoining buildings and environment as well as pipe string itself. Efficient jacking loads management thus enables long-distance pipejacking to be performed. The friction resistance usually constitutes the major component of jacking loads, which can be approximated from the minimum bound to the total jacking load and shows a cumulative nature while jacking in coarse ground. There have been many factors affecting the friction resistance and amongst the factors, lubricant plays a crucial role in managing the friction resistance. Lubricant should be maintained

H. El-Naggar et al. (Eds.): GeoMEast 2019, SUCI, pp. 121–134, 2020.
https://doi.org/10.1007/978-3-030-34252-4_10

within overcut annulus, and the percentage reduction of frictional stress appears to be linked to injected lubricant volumes. Cui et al. (2015) reported that due to loss of lubrication fluid into surrounding fissures, the effectiveness of slippery film was decreased leading that the jacking force of Line 2 of a pipejacking project conveying water from southern to northern Jiangsu was much higher than the three other lines (Lines 1 and 3–4). Pellet-Beaucour and Kastner (2002) indicated that for volumes varying from 25 to 170 L/m, frictional stress is reduced from 45 to 90%. However, the distribution characteristics of injected lubricant and their effects on the reduction of the friction resistance are rarely discussed.

The objectives of this study are (i) to present the distribution characteristics of lubricant injected with reference to the presented pipe ramming case history, (ii) to evaluate the lubrication performance of pipejacking in soft alluvial deposits using the baseline technique, and (iii) to investigate the impacts of the distribution characteristics of lubricant injected, soil and lubrication natures and misalignment on the lubrication performance.

2 Methodology

2.1 Distribution Characteristics of Injected Lubricant

The injection of lubricant can be categorised into three types which are single-point injection, two-point injection, and multi-point injection, and is linked to distribution characteristics while pipe ramming. The injection into the overcut annulus though a single injector is termed single-point injection, as depicted in Fig. 1a. Assuming the travel distance of injected lubricant is within 1 m extent, the overcut annulus at 4 m distance from the excavation face, while ramming RCP 7, is saturated through Injector 1 by injecting 0.27 m^3 lubricant. The injection into the overcut annulus by two injectors at two different distances from the face is termed two-point injection, as depicted in Fig. 1b. Injector 2 at a distance more close to the face would work with Injector 1 when the friction during pipe ramming gets higher than a single injector can manage. While jacking RCP 14, the overcut annulus at 4 m and 11 m distances from the face is saturated through Injectors 2 and 1 by injecting 0.13 m^3 (20.9%) and 0.50 m^3 (79.1%) lubricant, respectively. Effective lubrication cannot be sustained for long-distance pipe ramming. In this case multi-point injection may be used preventing inadequate lubrication from occurring, as depicted in Fig. 1c. While ramming RCP 23, a total of three injector is utilised, and the overcut annulus at 4 m, 14 m, and 21 m distances is saturated with 0.01 m^3 (5.5%), 0.07 m^3 (31.5%), and 0.15 m^3 (63.0%) lubricant, respectively.

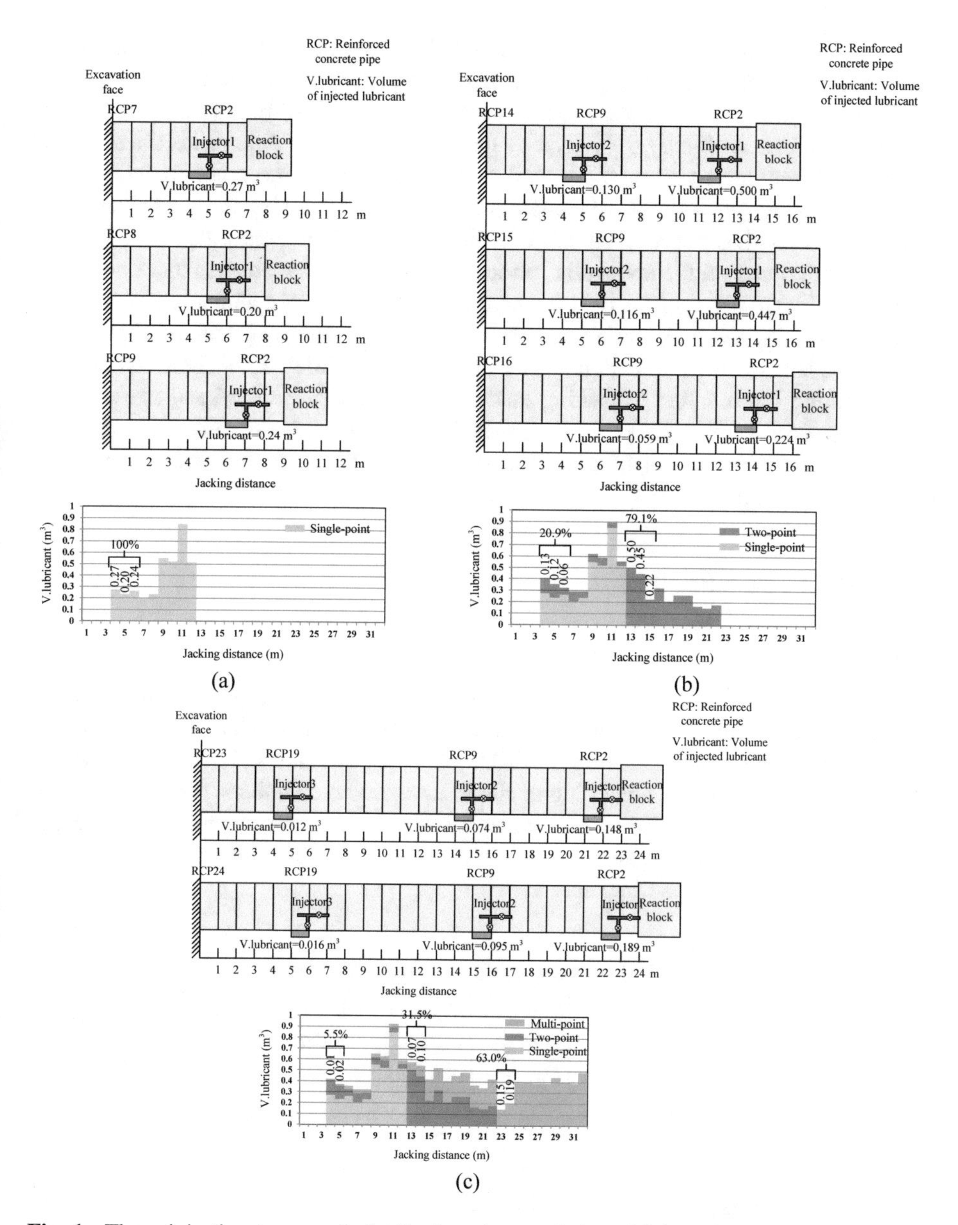

Fig. 1. Three injection types and distribution characteristics of injected lubricant: (a) single-point injection, (b) two-point injection and (c) multi-point injection

2.2 Lubrication Performance

Lubrication can largely affect the performance during pipe ramming, which can be assessed through the reduction of the frictional coefficient μ. Pellet-Beaucour and Kastner (2002) indicated that local variations of total jacking load are generally linked to the varying face resistance. A line constituted by the minima of total jacking load is thus referred to as baseline of jacking loads. The jacking forces from both ends of a baseline section divided by the section length leads to the average jacking force f_{avg}. The average jacking force f_{avg} divided by the circumference of pipe yields the frictional stress τ_{ld}. There are many methods available for calculating the pressure acting upon the pipe's crown (termed normal contact pressure σ_z hereafter) but the consensus about their correctness has not reached yet. Dividing the τ_{ld} value by the calculated σ_z value gives the frictional coefficient μ_{ld} for lubricated drives. The percentage reduction can be calculated by $(\mu_{nd} - \mu_{ld})/\mu_{nd} \times 100$ where the value of μ_{nd} is recommended by Stein et al. (1989) for soil-pipe (unlubricated) interface. The described method that can be used to evaluate the lubrication performance can be briefed in short as follows: (1) preparing the baseline of jacking loads, (2) calculating the frictional stress τ_{ld} for lubricated drives, (3) calculating the normal contact pressure σ_z, (4) evaluating the frictional coefficient μ_{ld} for lubricated drives, and (5) assessing the percentage reduction in the μ value.

3 Pipejacking Project

3.1 Background

The slurry shield was adopted to perform excavation of the two pipejacking drives in soft alluvial deposits. The alignment of all the drives is straight. During pipe ramming, the bentonite slurry with unit weight of 10.6 kN/m^3 was used to stabilise the excavation face and also transported tunnelling spoils to decantation chambers. Since the tunnels were excavated via the 1500-mm diameter cutter wheel, an overcut annulus of 30 mm was formed by using the smaller concrete pipe of 1440 mm in diameter. The two drives characteristics are detailed in Table 1.

Table 1. Characteristics of pipejacking drives

Parameters	Length (m)	Depth (m)	Pipe dia. D_e (m)	Cutter wheel dia. (m)	Soil cover h (m)	Face resistance (kPa)	V.inj (litre/m)
Drive C	75	10.8	1.44	1.5	10.1	555	552
Drive D	102	10.8	1.44	1.5	10.1	555	534

Note: V.inj = average volume of injected lubricant

3.2 Engineering Geology

The stratigraphic profile is established with reference to five 15-m deep geological boreholes penetrating through the 6-m thick silty sand layer into the 7.5 m thick poorly-graded to well graded sand and gravel layer. Based upon the standard penetration test and triaxial test results, the soil physical and mechanical properties are detailed in Table 2.

Table 2. Summary of soil physical and mechanical properties

Layer	Thickness (m)	SPT-N	γ (kN/m^3)	c' (kPa)	ϕ' (°)	q_u (kPa)
Backfill	2.0		17.6			
Silty sand with gravel	2.8	2–9	18.1	0	28	
Gravel and sand with cobble	>4.5	>100	20.1	0	35	6.7e3
Silty sand	>3.0	9–18	19.1	0	30	

4 Analysis and Discussions

4.1 Development of Jacking Loads Baseline

At Drive C, the shield spanned at a depth of 10.8 m and the associated activities are presented in Fig. 2a. The baseline of jacking forces consisted of the first 2–8 m section (gravel), 8–21 m (gravel) and 21–40 m (gravel) sections, and final 40–75 m (clayey

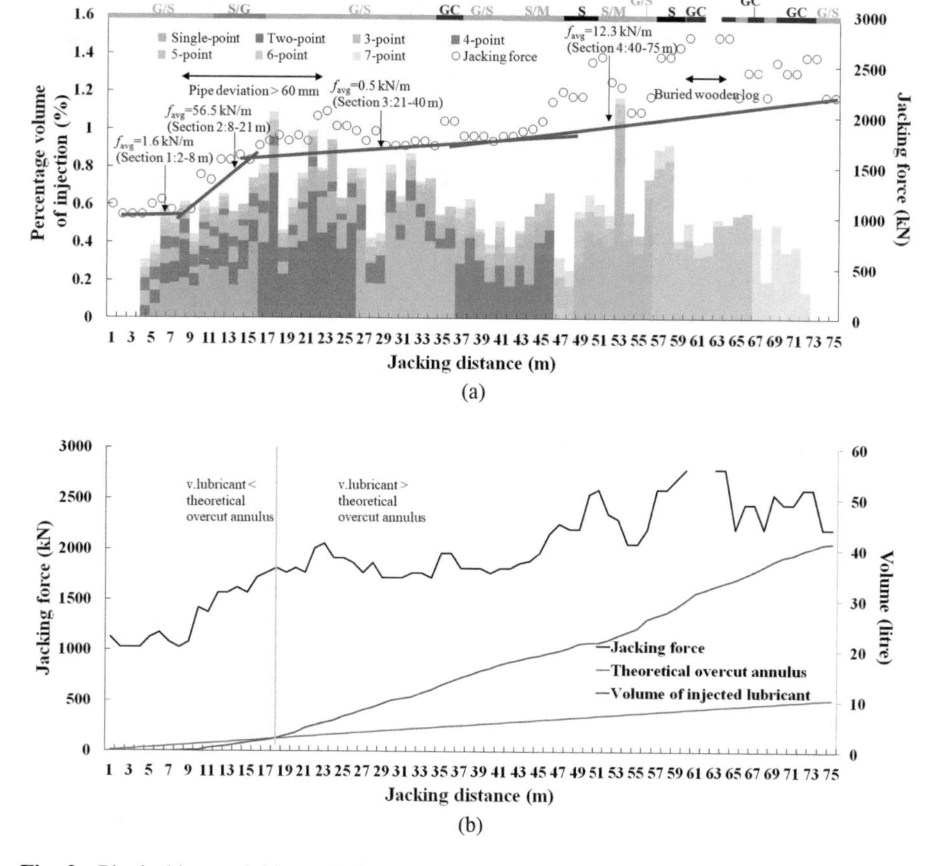

Fig. 2. Pipejacking activities at Drive C and the relationship between injection volume and jacking distance

gravel) section was established through the minima of jacking forces, with the average jacking forces f_{avg} at 1.6, 56.5, 0.5, and 12.3 kN/m, respectively. While their frictional stresses τ_{ld} were calculated as being 0.4, 12.5, 0.1, and 2.7 kPa, respectively.

The pipe ramming of Drive D traversed at a depth of 10.8 m and its activities are presented in Fig. 3a. The pipejacking results determined the baseline of jacking forces corresponding to the first 5 m (gravel) section (from 6 to 11 m), the subsequent 11–24 m (clayey gravel) and 24–31 m (clayey gravel) sections, and the final 31–102 m (clayey gravel) section, with the average jacking forces f_{avg} at 2.0, 18.1, 49.0, and 3.5 kN/m, respectively. The f_{avg} values divided by the circumference of pipe yielded the frictional stresses τ_{ld} equal to 0.4, 4.0, 10.8, and 0.8 kPa, respectively.

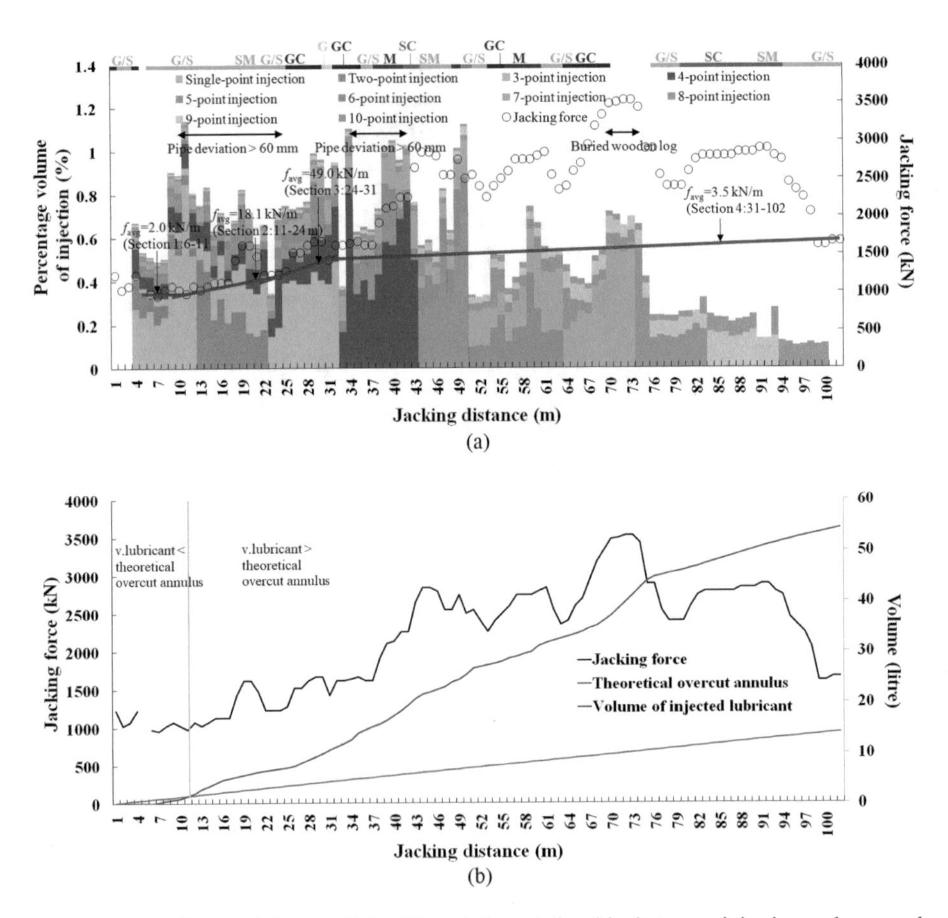

Fig. 3. Pipejacking activities at Drive D and the relationship between injection volume and jacking distance

4.2 Evaluation of Lubrication Performance

JMTA (JMTA 2013), ATV A 161 (German ATV rules and standards 1990), BS EN 1594 (BS 2009), ASTM F 1962 (2011), and GB 50332 (GB 50332-02 2002) are deemed to be the well-developed models that can be used to calculate the normal contact pressure σ_z. The models were modified with reference to the Terzaghi arching model founded upon active trap-door experiment (Terzaghi 1936) where the shear bands arise from the outside of tunnel cross sections along oblique lines, with a horizon included angle equal to 45° + $\phi/2$, and then they turn to vertical lines after passing the tunnel crown's level and finally arrive at the ground surface, as depicted in Fig. 4. The equation in the modified models was derived from the limit equilibrium of a horizontal slide (Terzaghi 1943):

$$dw - d\sigma_z - 2\tau_f = 0 \tag{1}$$

$$dw - d\sigma_z - 2(K\sigma_z \tan\phi + c) = 0 \tag{2}$$

$$\gamma B_1 dz - B_1 d\sigma_z - 2c dz - 2K\sigma_z \tan\phi dz = 0 \tag{3}$$

$$\frac{d\sigma_z}{dz} = \gamma - \frac{2c}{B_1} - \frac{2K\sigma_z \tan\phi}{B_1} \tag{4}$$

where B_1 = silo width, γ = unit weight of soil, K = soil pressure ratio, c = soil cohesion, τ_f = shear strength of soil, and σ_z = normal contact pressure. Equation (4) is a single order ordinary differential equation. Integrating Eq. (4) and considering boundary condition $\sigma_{z=0}$ = q at the surface deduces the σ_z value at any level:

$$\sigma_z = \frac{\gamma B_1 - 2c}{2Ktan\phi}\left(1 - e^{-2Ktan\phi\frac{z}{B_1}}\right) + qe^{-2Ktan\phi\frac{z}{B_1}} \tag{5}$$

In fact, the σ_z value should be calculated one stratum by one stratum from the surface to the bottom. Equation (5) can thus be rewritten as Eq. (6) for calculation of the σ_z value in multistratum environment.

$$\sigma_z(h_i) = \frac{\gamma_i B_1 - 2c_i}{2K_i tan\phi_i}\left(1 - e^{-2K_i tan\phi_i\frac{h_i}{B_1}}\right) + \sigma_z(h_{i-1})e^{-2K_i tan\phi_i\frac{h_i}{B_1}} \tag{6}$$

where i = stratum numbering from 1, 2, to n, h_i = thickness of ith stratum. Parameters used in the modified models are summarised in Table 3. The modified models assume the fully developed shearing bands. Given that the stratigraphy simplified from Zhang et al. (2016) and associated soil properties (Tables 4 and 5), comparisons of the normal contact pressure σ_z on the 17th pipe and the 24th pipe, respectively, between Terzaghi, JMTA, ATV A 161, BS EN 1594, ASTM F 1962, GB 50332, and the measured result were conducted (Fig. 5). It was evident that ATV A 161 and ASTM F 1962 give

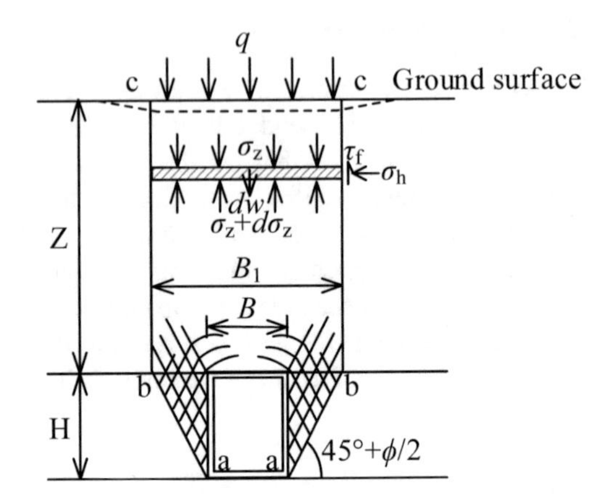

Fig. 4. Schematical illustration of Terzaghi arching model (after Terzaghi 1943)

estimations in good agreement with field measurements for jacked pipes. Thus, ATV A 161 and ASTM F 1962 were selected as the preferred models for calculating the normal contact pressure σ_z. Dividing the frictional stress τ_{ld} by the calculated normal contact pressure σ_z led to the frictional coefficient μ_{ld}. The percentage reduction was obtained using $(\mu_{nd} - \mu_{ld})/\mu_{nd} \times 100$, as listed in Table 6.

Table 3. Parameters used in the modified models

Parameters	Silo width b (m)	Friction angle in shear plane δ (°)	Soil pressure ratio K	Cohesion c (kPa)
Terzaghi (1943)	$D_e \times [1 + 2\tan\alpha]$	ϕ	1	c
JMTA (2013)	$(D_e + 0.08) \times (\sec\alpha + \tan\alpha)$	ϕ	1	c
ATV A 161 (1990)	$1.732D_e$ ($\phi = 30°$)	$\phi/2$	$K_0 = 0.5$	None
BS EN 1594 (2009)	$D_e \times (1 + 2\tan\alpha)$	ϕ	K_0	c (with verification)
ASTM F 1962 (2011)	$1.5D_e$	$\phi/2$	$\tan^2(45° - \phi/2)$	None
GB 50332 (2002)	$D_e \times (1 + \tan\alpha)$	$K_a\tan\phi = 0.19$		None

Note: $\alpha = 45° - \phi/2$; D_e = outer pipe diameter; D_b = tunnel bore diameter; ϕ = soil friction angle; K = soil pressure ratio; K_0 = soil pressure ratio at rest; c = soil cohesion

Table 4. Soil properties for 17th pipe (after Zhang et al. 2016)

Parameters for 17th pipe	Thickness (m)	γ (kN/m^3)	c' (kPa)	ϕ' (°)
Medium to coarse sand with clay	9.3	17.6	0	30.2
Silty clay	17.6	18.3	10.6	6.9

Table 5. Soil properties for 24th pipe (after Zhang et al. 2016)

Parameters for 24th pipe	Thickness (m)	γ (kN/m^3)	c' (kPa)	ϕ' (°)
Medium sand with peat	8.3	17.2	0	30.6
Silty clay	2.0	17.6	10.6	6.7
Coarse sand with clay	5.3	18.6	0	32.1
Silty clay	6.7	18.1	10.6	6.7
Fine to medium sand with clay	2.2	18.6	0	31.5

Table 6. Summary of reduction in μ against each baseline section

	Avg. jacking force (kN/m)	τ_{ld} (kPa)	σ_z by ATV A 161 (kPa)	μ_{ld}	Red.μ (%)
Drive C	Section 1: 1.6 Section 2: 56.5 Section 3: 0.5 Section 4: 12.3	Section 1: 0.4 Section 2: 12.5 Section 3: 0.1 Section 4: 2.7	Section 1: 111.1 Section 2: 111.1 Section 3: 111.1 Section 4: 102.0	Section 1: 0.004 Section 2: 0.1 Section 3: 0.0009 Section 4: 0.03	Section 1: 98 Section 2: 71 Section 3: 99 Section 4: 88
Drive D	Section 1: 2.0 Section 2: 18.1 Section 3: 49.0 Section 4: 3.5	Section 1: 0.4 Section 2: 4.0 Section 3: 10.8 Section 4: 0.8	Section 1: 111.1 Section 2: 102.0 Section 3: 102.0 Section 4: 102.0	Section 1: 0.004 Section 2: 0.04 Section 3: 0.1 Section 4: 0.008	Section 1: 98 Section 2: 84 Section 3: 60 Section 4: 97

Note: τ_{ld} = frictional stress; σ_z = calculated normal contact pressure; μ_{ld} = backanalysed frictional coefficient; Red.μ = percentage reduction in μ value

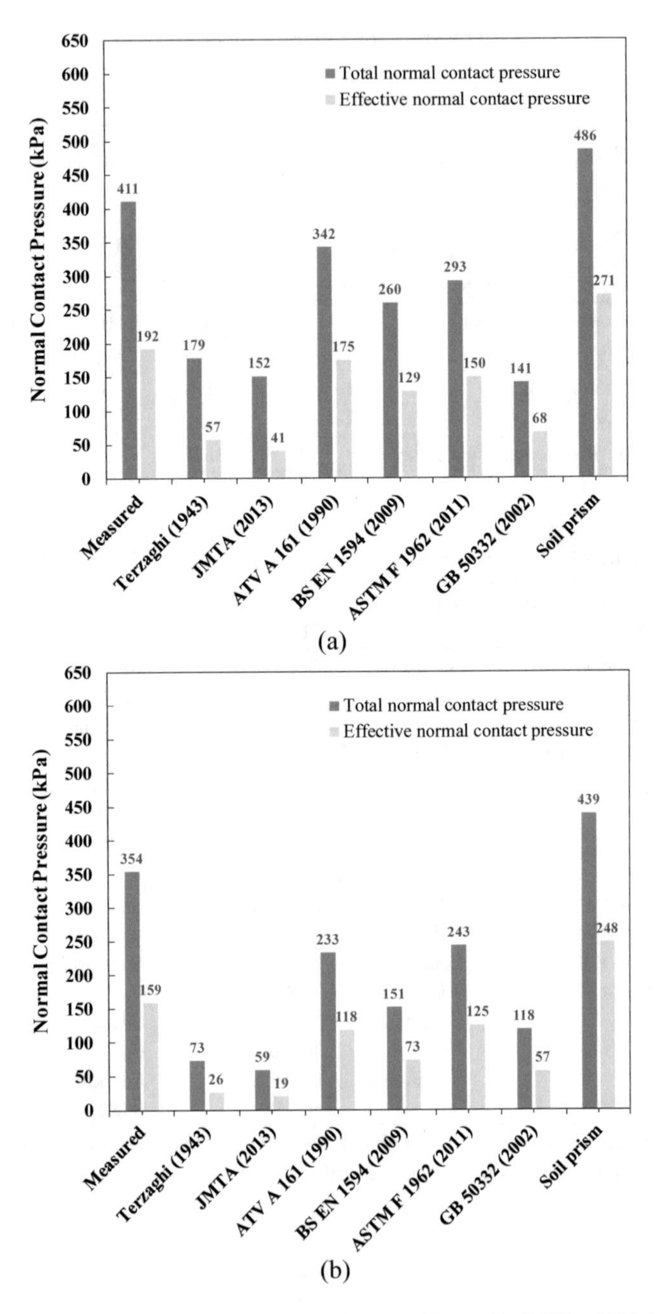

Fig. 5. Comparisons of normal contact pressure between Terzaghi, JMTA, ATV A 161, BS EN 1594, ASTM F 1962, GB 50332, and the measured result: (a) 17th pipe and (b) 24th pipe

4.3 Effect of Lubricant Injection Mode

Two drives (Drives C and D) were analysed because of their good data completeness. The cumulative volumes and associated percentages for Drives C and D were estimated relying only upon lubricant injections into overcut annulus of the analysed baseline section (Tables 7 and 8).

Table 7. Summary of cumulative volume at Drive C

Injector location	RCP 02	RCP 09	RCP 19	RCP 29	RCP 38	RCP 48	RCP 58
Section 1 (2–8 m)	0.84 (100)						
Section 2 (8–21 m)	4.8 (79.2)	1.26 (20.8)					
Section 3 (21–40 m)	7.06 (60)	2.35 (20)	1.18 (10)	1.18 (10)			
Section 4 (40–75 m)	9.08 (40)	4.54 (20)	2.73 (12)	2.73 (12)	1.82 (8)	0.91 (4)	0.91 (4)

Note: Number in bracket indicates the percentage volume of injected lubricant

Table 8. Summary of cumulative volume at Drive D

Injector location	RCP 02	RCP 09	RCP 19	RCP 29	RCP 38	RCP 48	RCP 58
Section 1 (6–11 m)	4.05 (100)						
Section 2 (11–24 m)	2.90 (63)	1.45 (31.5)	0.25 (5.5)				
Section 3 (24–31 m)	3.34 (63)	1.67 (31.5)	0.29 (5.5)				
Section 4 (31–102 m)	14.67 (36.2)	8.79 (21.7)	3.48 (8.6)	3.48 (8.6)	3.48 (8.6)	1.66 (4.1)	1.66 (4.1)
Injector location	RCP 68	RCP 78	RCP 88				
Section 1 (6–11 m)							
Section 2 (11–24 m)							
Section 3 (24–1 m)							
Section 4 (31–102 m)	1.66 (4.1)	0.81 (2)	0.81 (2)				

Note: Number in bracket indicates the percentage volume of injected lubricant

A 30 mm over excavation outside the 1.44-m diameter pipe justified the overcut ratio of 0.02 corresponding to the theoretical overcut annulus of 0.138 m^3/m. The injection mode for the four sections of gravel (first three sections at Drive C and first section of at Drive D) refers to Tables 6 and 7. Their injection volumes averaged 0.139 m^3/m (0.84 m^3/6 m), 0.466 m^3/m (6.06 m^3/13 m), 0.619 m^3/m (11.77 m^3/19 m), and 0.811 m^3/m (4.05 m^3/5 m), respectively, and were greater than 0.138 m^3/m. The excessive injection volumes not only indicated a strong intention to saturate the overcut annulus to sustain effective lubrication conditions, but also showed a significant loss of lubricant while pipe ramming. Their percentage reductions in the μ value from ATV A 161 were calculated as being 98% (0.004 vs. 0.35), 71% (0.1 vs. 0.35), 99% (0.0009 vs. 0.35), and 98% (0.004 vs. 0.35), respectively. The buoyancy of 17.3 kN greater than the pipe self-weight of 12.6 kN and the saturated overcut made a 71% reduction (the second lowest in this study) from 0.35 (the average μ value recommended by Stein et al. (1989) for gravel-pipe interface) to 0.1. The major cause to lead to the second lowest reduction was attributed to the excessive pipe deviation (Fig. 2a). The μ_{ld} value of 0.1 matched the lower limit recommended by Stein et al. (1989) for lubricated drives, suggesting that the 8–21 m section through gravel was well-lubricated. Also, the overcut full of lubricant and the sufficient buoyancy made the reductions high enough for the other three sections, with the μ_{ld} values smaller than 0.1, indicating that lubrication during pipe ramming of the other three sections was very effective.

Tables 6 and 7 details the injection mode for the other four sections of clayey gravel (final section at Drive C and subsequent three sections at Drive D). Their injection volumes averaging 0.649 m^3/m (22.72 m^3/35 m), 0.354 m^3/m (4.60 m^3/13 m), 0.757 m^3/m (5.30 m^3/7 m), and 0.570 m^3/m (40.5 m^3/71 m), respectively, were also in excess of the theoretical overcut annulus of 0.138 m^3/m. The excessive injection volumes made the percentage reductions to reach to 88% (0.03 vs. 0.25), 84% (0.04 vs. 0.25), 60% (0.1 vs. 0.25), and 97% (0.008 vs. 0.25), respectively. The sufficient buoyancy and the lubricant-saturated overcut caused an 84% reduction from 0.25 suggested by Stein et al. (1989) for clay-pipe interface to 0.04. The varying face resistance induced by jacking into the clayey gravel and the excessive pipe deviation were deemed to be the major cause to lead to the reduction of 84%. The μ_{ld} value equal to 0.04 was far less than the lower limit, which also indicated adequate lubrication. It is worth to note that jacking into gravel at 31 m distance caused the reduction of 60% (the lowest in this study). This phenomenon was most likely because of the gravel not being long enough to develop lower face resistance. The use of the excessive injection volumes accompanied with the justified buoyancy made the percentage reduction far less than 0.1 for the other two sections, which also indicated adequate lubrication. To short, the effect of misalignment and the varying face resistance contributed to the low percentage reductions despite the excessive injection volumes. The occasional gravel led to the misleading reduction. The effects outweighed the influence of injection mode on the lubrication performance.

4.4 Effect of Soil and Lubricant Natures

The permeation of lubricant into the surrounding geology would mitigate the effort of establishing a lubricating layer at soil-pipe interface. This is most likely because of the inability of the lubricant to develop a filter cake of low permeability. Such a permeable overcut could also result in injection volume in excess of the theoretical overcut. The phenomena discovered in this study resulted in the relatively large injection volume of 0.552 m^3/m at Drive C (including mostly the gravel) than 0.534 m^3/m at Drive D (including mostly the clayey gravel). On the other hand, the continuous injection of the lubricant with Marsh cone viscosity of 38 min into the overcut annulus effectively reduced the friction resistance to viscous resistance. This led the reductions in excess of 88%, which is consistent with Staheli et al. (2006).

4.5 Effect of Misalignment

The effect of misalignment significantly exaggerates the friction resistance and has implications on the lubrication performance. There was an increase of the jacking force of 771 kN while spanning between 8 and 21 m distance at Drive C. This increase in the jacking force caused the percentage reduction in the μ value to reduce by 27% from 98%. The leading cause was not because of the overcut not full of lubricant, but because of the exaggerating frictional stress τ_{ld} of 12.5 kPa induced by the pipe deviation in excess of 60 mm, which is almost 30 times larger than 0.4 kPa of the 2–8 m section in the same gravel. It is worth to note from Fig. 2b that the overcut annulus was saturated after the pipe at 19 m distance was rammed, indicating that inadequate lubrication also contributed to the reduction of 71%. While traversing through the 11–24 m section of gravel at Drive D, the jacking force increased by 637 kN to 1617 kN. The increase of 637 kN reduced the percentage reduction to 84% from 98%, most likely because of the exaggerating frictional stress τ_{ld} of 4.0 kPa, which is 5 times larger than 0.8 kPa of the 31–102 m section in the same clayey gravel. The cause to lead to the reduction of 84% was attributed to the combined effects of misalignment and change in the face resistance, resulting from pipe ramming into the clayey gravel.

5 Conclusions

In this study, the method that can be utilised for assessing the lubrication performance using the percentage reduction in the frictional coefficient was described. The effect of lubricant injection mode, soil and lubricant natures, and misalignment on the lubrication performance was investigated. Based upon the analysis and discussions, the following conclusions can be drawn:

(1) The excessive pipe deviation and the change in the face resistance, resulting from pipe ramming into the clayey gravel, contributed to the low percentage reductions despite the excessive injection volumes. The gravel that presented occasionally led to the inability to develop lower face resistance and caused the misleading percentage reduction of 60%. The effects outweighed the influence of injection mode on the lubrication performance.

(2) The permeation of lubricant into the surrounding geology was due to the inability of the lubricant to develop a filter cake of low permeability. Such a permeable overcut also resulted in the injection volumes in excess of the theoretical overcut of 0.138 m^3/m. The phenomena discovered in this study justified the relatively large injection volume of 0.552 m^3/m at Drive C (including mostly the gravel) than 0.534 m^3/m at Drive D (including mostly the clayey gravel).
(3) The increase of 771 kN in the jacking force caused the percentage reduction to reduce by 27% from 98%. This was attributed to the exaggerating frictional stress τ_{ld} of 12.5 kPa induced by the pipe deviation in excess of 60 mm, which is almost 30 times larger than 0.4 kPa of the 2–8 m section in the same gravel. The inadequate lubrication also contributed to the reduction of 71%.

References

Cui, Q.L., Xu, Y.S., Shen, S.L., Yin, Z.Y., Horpibulsuk, S.: Field performance of concrete pipes during jacking in cemented sandy silt. Tunneling Undergr. Space Technol. **49**, 336–344 (2015). https://doi.org/10.1016/j.tust.2015.05.005

Pellet-Beacour, A.L., Kastner, R.: Experimental and analytical study of friction forces during microtunneling operations. Tunn. Undergr. Space Technol. **17**(1), 83–97 (2002)

Japan Microtunnelling Association (JMTA): Mictotunnelling Methods Serious II, Design, Construction Management and Rudiments, Japan Microtunnelling Association (JMTA), Tokyo, pp. 69–72 (2013)

German ATV rules and standards: ATV-A 161 E-90. Structural calculation of driven pipes. Hennef, pp. 18–20 (1990)

British standards: BS EN:1594-09 Gas supply system-pipelines for maximum operating pressure over 16 bar-functional requirements. Brussels, pp. 76–78 (2009)

ASTM: Standard Guide for Use of Maxi-Horizontal Directional Drilling for Placement of Polyethylene Pipe or Conduit under Obstacles Including River Crossings. F 1962-11, West Conshohocken, PA (2011)

The Ministry of Construction of China: GB 50332-02, Structural design code for pipeline of water supply and waste water engineering. Beijing, pp. 11–12 (2002)

Terzaghi, K.: The shearing resistance of saturated soils and the angle between the planes of shear. In: Proceedings of the 1st International Conference on Soil Mechanics and Foundation Engineering. Harvard University Press, Cambridge, MA, pp. 54–56 (1936)

Terzaghi, K.: Theoretical Soil Mechanics, pp. 66–76. Wiley, New York (1943)

Stein, D., Möllers, K., Bielecki, R.: Microtunneling: Installation and Renewal of Nonman-Size Supply and Sewage Lines by the Trenchless Construction Method. Ernst, Berlin (1989)

Staheli, K.: Jacking force prediction an interface friction approach based on pipe surface roughness. Dissertation, Georgia Institute of Technology (2006)

Effect of Soil–Structure Interaction on Free Vibration Characteristics of Antenna Structure

Venkata Lakshmi Gullapalli[1(✉)], Neelima Satyam[2], and G. R. Reddy[3]

[1] HBNI, SM, ECIL, Hyderabad, India
gullapalli30@ecil.co.in
[2] IIT, Indore, India
neelima.satyam@iiti.ac.in
[3] SSES, BARC, HBNI, Maharashtra, India
rssred@barc.gov.in

Abstract. The influence of **soil-structure interaction** in the Natural Frequency of a parabolic reflector antenna is investigated. The main free vibration characteristic of an antenna is natural frequency. The analysis model simulated with fixed base condition and soil structure interaction. A parametric study is conducted to understand the antenna structure soil behavior (soil structure interaction (SSI)) by changing the antenna orientation, soil properties with impedance functions (spring constants) as per ASCE 4-16 and 3D (three dimensional modeling of founding medium) model methods. To address this problem, Finite Element Method is used to model soil structure interaction analysis of antenna structure with foundation by MSC PATRAN & NASTRAN using direct method and impedance method. The soil considered as homogeneous and analyzed for shear wave velocity (Vs) 100 m/s, 180 m/s, 250 m/s, 450 m/s, 760 m/s for antenna with 90° elevation. To study the embedment effect, soil around the footing and pedestal is considered. The frequency results are compared for antenna with four orientations with Vs 100 m/s soil properties. Considerable reduction in frequency observed with soil structure interaction in higher modes. Soil structure interaction study in antennas helps in finding what extent the natural frequency falls when the soil interaction is incorporated in the model.

Keywords: Normal modes · Antenna Structure · Shear wave velocity · Natural frequency · Spring constant

1 Introduction

Establishing of parabolic reflector antennas is required for any satellite communications. All ground station antennas are normally installed on the ground or on top of the building based on RF requirements. The important parameters of antenna structure are Gain, Pointing accuracy and Natural frequency. These parameters are depending on antenna stiffness, loads acting on that and sub soil conditions. Most of the literature on **soil structure interaction** available for buildings, towers, chimneys. However, for parabolic reflector ground station antenna structures, published literature not available. The soil structure interaction effect on dynamic behavior of antenna structure is important to get effective and satisfactory performance of antenna.

H. El-Naggar et al. (Eds.): GeoMEast 2019, SUCI, pp. 135–144, 2020.
https://doi.org/10.1007/978-3-030-34252-4_11

Ground station parabolic reflector antennas are classified as Limited motion, full coverage, full motion depending on satellite tracking requirements. Limited motion means, the antenna will move azimuth and elevation within the limits i.e., for example ±60° in azimuth and Full coverage antenna means it covers azimuth ±360° and elevation 0 to 90°. Full motion antenna means antenna is in continuous rotation ±360° in azimuth and ±90° in elevation with given velocity. For antennas mounting on loose soils with shear wave velocity in the range of 100 m/s to 180 m/s, the effect of **soil structure interaction** is studied on 7.2 m diameter full coverage antenna.

For comparing the natural frequency of antenna structure, analysis model created MSC PATRAN and imported to NASTRAN solver for response estimation. Results of fixed base analysis and Soil Structure Interaction analysis are compared. The soil parameters are varied with shear wave velocity as 100 m/s, 180 m/s, 250 m/s, 450 m/s and 760 m/s with homogeneous properties for Antenna with 90° elevation orientation (Zenith) i.e. looking to sky. Results for the four orientations (i.e., 0, 45, 74, 90°) are compared with soil property corresponding to Vs 100 m/s.

2 Description of 7.2 m Dia Full Coverage Antenna

The full coverage antenna structure consists of parabolic main reflector with central opening for mounting feed on hub, sub-reflector, quadripod, backup structure, hub, yoke arm and screw jack for elevation rotation, azimuth bearing and pedestal. The main reflector consists of 16 panels made of paraboloid structure. Backup structure consists of 16 radial trusses connecting with circumferential members. Hub is of cylindrical shell made of aluminium. Pedestal consists of two parts, bottom is cylindrical, top one

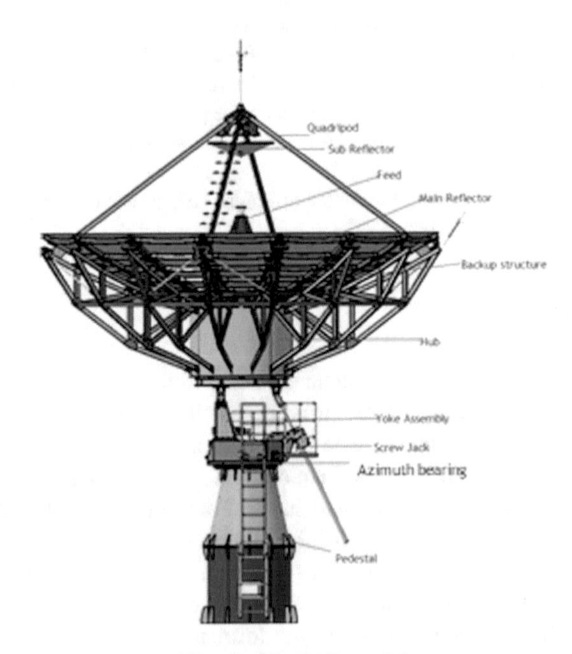

Fig. 1. 3D CAD model

is circular tapered cylinder. The reflector, sub reflector, backup structure, quadripod are made of aluminium. Yoke arm, pedestal made of steel. The 3D CAD model given in Fig. 1 describes the antenna structure. Antenna structure is mounted on isolated RCC foundation through anchor bolts.

The RCC foundation is square in plan of size 4 m × 4 m × 0.6 m raft and 2.1 × 2.1 × 1.7 m pedestal.

3 FEM Model

Quadripod, backup structure, panel stiffeners, screw rod are modeled as 1-D bar-/-beam elements. Main reflector, hub, pedestal and yoke arms are modeled as shell elements (Quad-4). Foundation and soil are modelled as solid elements (Hex-8). Sub-reflector, feed elements are created as lumped mass elements. Hub to yoke arm, pedestal to Foundation, feed mounting created using multi point constraints RBE-2 elements.

Analysis results are obtained and compared for 3 cases. Case 1 is with fixed base, Case 2 is with impedance functions i.e., with soil stiffness (lumped representation of soil structure interaction), Case 3 is with direct method, where soil is modeled as per ASCE 4-98 clause 3.3.3 [1]. For Case 3, soil boundary is considered up to 2.5 times the size of footing in plan and 1.5 times the size of footing in depth for soil analysis [1, 3] (Tables 1 and 2).

Table 1. Material properties

Property	Materials properties used for modeling		
	Aluminum	Steel	Concrete (M20)
Density (kg.cu.m)	2750	7850	2500
Poisson's ratio	0.30	0.30	0.17
Young's modulus MPa	0.7×10^5	2.1×10^5	0.2236×10^5

Table 2. Soil properties for various shear wave velocities

Shear wave velocity V_s m/sec	Density of soil ρ Kg/m^3	Poisson's ratio μ	Shear Modulus $G = \rho\ Vs^2$ N/m^2	Shear Modulus $G = \rho\ Vs^2$ Mpa	Elastic Modulus $E = G\ (2\ (1 + \mu))$ in N/m^2	Elastic Modulus $E = G\ (2(1 + \mu))$ in Mpa
100	1980	0.25	19800000	19.80	49500000	49.50
170	1830	0.3	52890000	52.89	137510000	137.51
180	1830	0.3	59292000	59.29	154159200	154.16
250	1900	0.3	118750000	118.75	308750000	308.75
450	1900	0.3	384750000	384.75	1000350000	1000.35
760	2100	0.3	1212960000	1212.96	3153696000	3153.69

For soil structure interaction with Impedance functions, equivalent spring constants are calculated using Table 3.3.3 of ASCE 4-98 [1]. Only horizontal and vertical spring stiffness are considered for analysis (Table 3).

Table 3. Lumped representation of structure – soil interaction at surface for rectangular base

Motion	Equivalent spring constant
Horizontal	$k_x = k_y = 2(1 + \mu)\, G\beta_x\sqrt{BL}$
Vertical	$k_z = (G/(1 - \mu))\beta_z \sqrt{BL}$

Where B is width of basemat, L is length of basemat, μ is poisons ratio of foundation medium, G is shear modulus, β_x, β_z are constants depending on basemat dimensional ratio (as per Fig. 3.3.3 ASCE 4-98) [1], R is equivalent radius of basemat.

Spring constants are calculated for square base (Table 4).

Table 4. Equivalent spring constants for various soil properties

Shear wave velocity V_s m/sec	Square footing size in m B = L	β_x = constant function of dimensional ratio for square footing L/B = 1.0	Equivalent spring constant K horizontal for rectangular footing = $2 (1 + \mu) G\beta_x \sqrt{BL}$ MN/m	β_z = constant function of dimensional ratio for square footing L/B = 1.0	Equivalent spring constant K vertical for rectangular footing = $(G/(1 - \mu)) * \beta_z * \sqrt{BL}$ MN/m
100	4	0.9	178.20	2.2	232.32
170	4	0.9	495.02	2.2	664.86
180	4	0.9	554.97	2.2	745.38
250	4	0.9	1111.50	2.2	1492.85
450	4	0.9	3601.26	2.2	4836.86
760	4	0.9	11353.30	2.2	15248.64

4 Analysis

Parametric study carried out on A1 type 7.2 m diameter full coverage antenna with fixed base and with Soil structure Interaction. The study is with four orientations (i.e., 0, 45, 74 and 90 elevation), homogeneous and layered soil and with six shear wave velocities (100, 170, 180, 250, 450, 760 m/s). Both Fixed base and Soil Structure interaction analyses are carried out. SSI carried out with direct method and impedance method, for three cases as detailed below (Figs. 2 and 3).

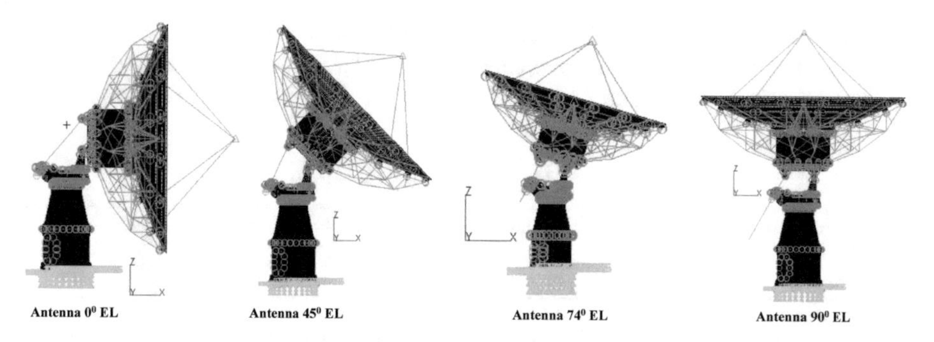

Fig. 2. Antenna structure FEM model

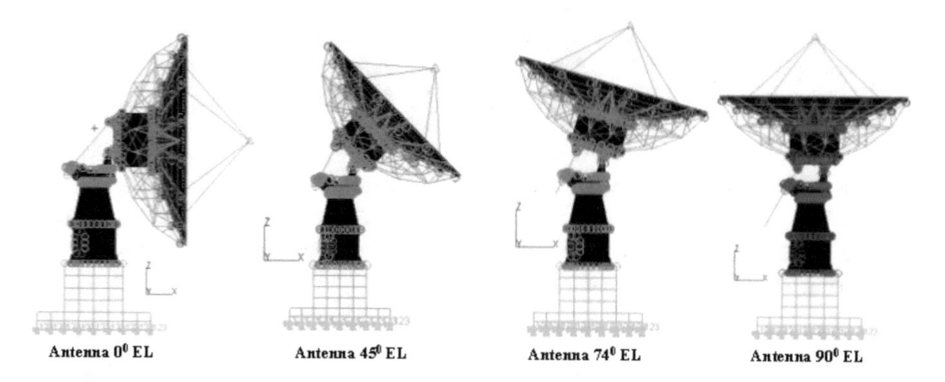

Fig. 3. Antenna structure+foundation FEM model

Case 1: Fixed Base Analysis

a. Antenna Structure with anchor bolts locations fixed (both translations and rotations).
b. Antenna Structure with Foundation. Translations fixed at raft bottom.

Case 2: Soil Structure Interaction – With Impedance Functions as per ASCE 4-16

a. Antenna Structure with Foundation. Only translational springs considered. The effect of rocking and torsion not considered.

Case 3: Soil Structure Interaction – Direct Method with 3D Soil

a. Antenna Structure, foundation and Soil. (translations fixed at bottom).

Antenna Structure Weight = 5.908 MT, Foundation Weight = 42.752 MT
Total weight acting on Soil is 48.66 MT

Modal Participation factors for antenna with 90° elevation are, 1st Mode is 23% (3.807 Hz bending in x) actively participated in X, 2nd mode is 25% (4.325 Hz bending in y) actively in Y 11th mode 24.5% (18.52 Hz) bending in z) in Z (Table 5).

Table 5. Fixed base analysis results

S. No.	Description of mode shape	Natural frequency in Hz antenna structure				Natural frequency in Hz antenna + foundation			
		El 0°	El 45°	El 74°	El 90°	El 0°	El 45°	El 74°	El 90°
1	Bending in x	3.606 1st mode	4.229 1st mode	4.058 1st mode	3.807 1st mode	3.605 1st mode	4.228 1st mode	4.05 1st mode	3.801 1st mode
2	Bending in y	4.262 2nd mode	4.36 2nd mode	4.36 2nd mode	4.325 2nd mode	4.256 2nd mode	4.359 2nd mode	4.35 2nd mode	4.317 2nd mode
3	Torsion in reflector	6.99 3rd mode	6.343 3rd mode	6.338 3rd mode	6.704 3rd mode	6.987 3rd mode	6.342 3rd mode	6.33 3rd mode	6.7 3rd mode
4	Bending in Z	14.612 6th mode	22.327 11th mode	18.717 9th mode	18.51 11th mode	14.61 6th mode	22.31 11th mode	18.68 9th mode	18.44 11th mode

Case 2: Soil Structure Interaction: Impedance Functions

Soil stiffness assigned at bottom of foundation raft. 3 translational (kx, ky, kz) and 3 rotation (k torsion) stiffness's assigned at middle, corner and edge as per influence area of raft (Fig. 4 and Tables 6 and 7).

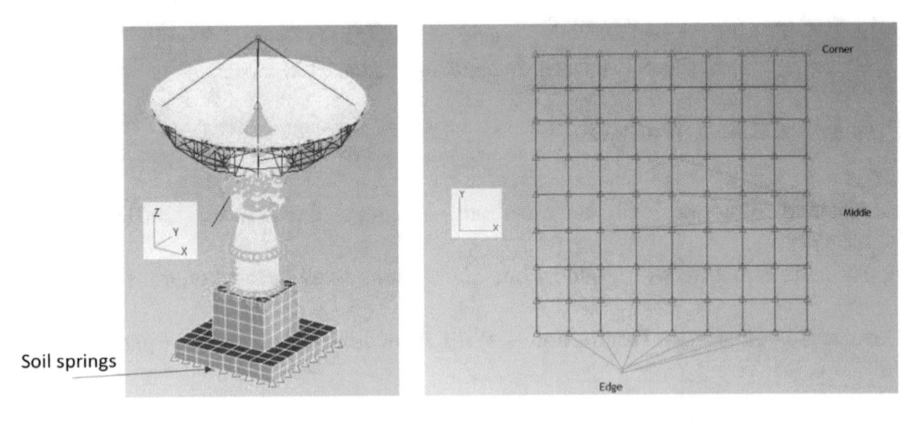

Fig. 4. Antenna structure+foundation with soil springs

Natural frequency is same as fixed base analysis for first 3 modes with higher shear wave velocity soil springs. For Vs 100 m/s, considerable reduction is observed for higher modes.

10.55 Hz obtained as bending in Z (vertical) with Vs 100 m/s. 18.324 Hz obtained for Vs 760 m/s.

Table 6. Analysis results with soil springs

	Description of mode shape	Natural frequency in Hz for Vs 100 m/s			
		Antenna in 0° EL	Antenna in 45° EL	Antenna in 74° EL	Antenna in 90° EL
1	Bending in x reflector	3.431 1st mode	3.5686 1st mode	3.415 1st mode	3.289 1st mode
2	Bending in y reflector	3.741 2nd mode	3.598 2nd mode	3.544 2nd mode	3.537 2nd mode
3	Bending in x all parts	5.067 3rd mode	6.1224 4th mode	6.395 4th mode	6.239 3rd mode
4	Torsion in reflector	7.722 5th mode	7.151 5th mode	6.162 3rd mode	6.529 4th mode
5	Bending in y all parts	5.518 4th mode	5.781 3rd mode	6.763 5th mode	6.683 5th mode
6	Bending in Z	10.794 6th Mode	10.687 6th mode	10.58 6th mode	10.55 6th mode

Table 7. Analysis results with soil springs

S. No.	Description of mode shape	Natural frequency in Hz for antenna with 90° elevation					
		Vs 100 m/s	Vs 170 m/s	Vs 180 m/s	Vs 250 m/s	Vs 450 m/s	Vs 760 m/s
1	Bending in x reflector	3.289 1st mode	3.636 1st mode	3.655 1st mode	3.729 1st mode	3.778 1st mode	3.792 1st mode
2	Bending in y in reflector	3.537 2nd mode	4.058 2nd mode	4.087 2nd mode	4.204 2nd mode	4.281 2nd mode	4.303 2nd mode
3	Torsion in reflector	6.529 4th mode	6.659 3rd mode	6.664 3rd mode	6.683 3rd mode	6.695 3rd mode	6.699 3rd mode
4	Bending in z	10.551 6th mode	16.414 11th mode	16.806 11th mode	17.722 11th mode	18.164 11th mode	18.324 11th mode

The significant mass participation is there with soil structure interaction. 41.8% (12.1 Hz 7th Mode) in X, 30.3% (6.68 Hz 5th Mode) in Y, 98.4% (10.6 Hz 6th Mode) in Z.

Case 3: Soil Structure Interaction: Direct Method

Direct method carried out with Vs 100 m/s and without embedment effect i.e. raft resting on soil, around the footing and rcc pedestal soil not considered for four orientations (Fig. 5).

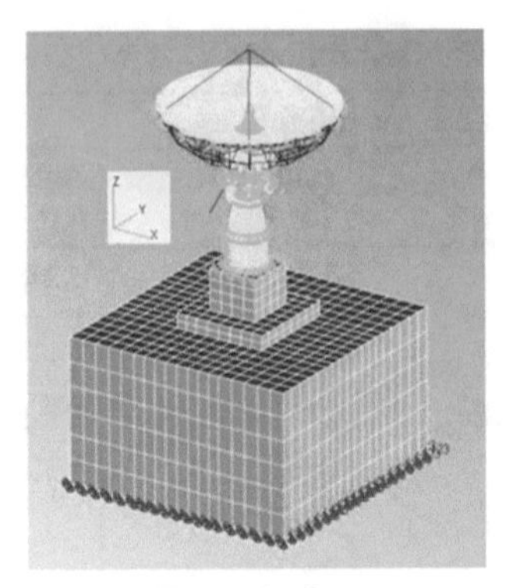

Foundation resting on soil

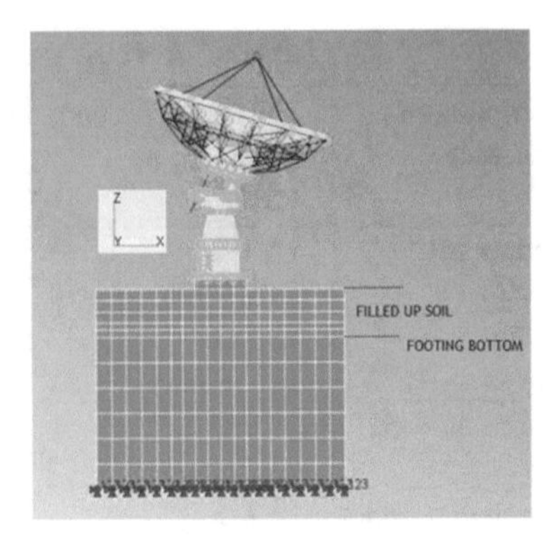

Foundation embedded in soil

Fig. 5. Soil structure interaction direct method model

Soil dynamic properties Shear modulus, Elastic modulus and poisons ration only defined. Soil density not assigned to soil elements (Tables 8 and 9).

Table 8. Analysis results with raft resting on soil

	Description of mode shape	Natural frequency in Hz for Vs 100 m/s			
		Antenna in 0° EL	Antenna in 45° EL	Antenna in 74° EL	Antenna in 90° EL
1	Bending in x reflector	3.564 1st mode	4.025 1st mode	3.841 1st mode	3.635 1st mode
2	Bending in y reflector	4.101 2nd mode	4.106 3rd mode	4.07 2nd mode	4.049 2nd mode
3	Torsion mode in reflector	6.430 3rd mode	6.165 3rd mode	6.292 3rd mode	6.663 3rd mode
4	Bending in z vertical	14.664 9th mode	21.559 14th mode	19.297 12th mode	18.924 14th mode

Table 9. Analysis results with foundation embedded in soil

	Description of mode shape	Natural frequency in Hz for Vs 100 m/s			
		Antenna in 0° EL	Antenna in 45° EL	Antenna in 74° EL	Antenna in 90° EL
1	Bending in x reflector	3.584 1st mode	4.121 1st mode	3.919 1st mode	3.7101 1st mode
2	Bending in y reflector	4.173 2nd mode	4.221 2nd mode	4.183 2nd mode	4.171 2nd mode
3	Torsion in reflector	6.741 3rd mode	6.251 3rd mode	6.249 3rd mode	6.681 3rd mode
9	Bending in Z	14.561 8th mode	22.212 14th mode	19.545 11th mode	14.699 10th mode

The significant mass participation is there with soil structure interaction. 82.6% (8.28 Hz) in X, 81.5% (8.39 Hz) in Y, 76.8% (10th Mode 14.7 Hz) in Z (Tables 10 and 11).

Table 10. Comparison of analysis results fixed base, SSI (soil springs and direct method) for 90° elevation orientation

Description of mode	Fixed base (antenna + foundation)	With Vs 100 m/s soil springs	Foundation resting on soil without embedment with Vs 100 m/s	Foundation surrounded by soil with embedment with Vs 100 m/s
Bending in X reflector	3.801 1st mode	3.289 1st mode	3.6353 1st mode	3.711 1st mode
Bending in y reflector	4.317 2nd mode	3.537 2nd mode	4.049 2nd mode	4.171 2nd mode
Torsion mode in reflector	6.699 3rd mode	6.529 4th mode	6.663 3rd mode	6.6808 3rd mode
Bending in z vertical	18.441 11th mode	10.551 6th mode	18.924 14th mode	14.699 10th mode

Table 11. Comparison of mass participation fixed base, SSI (soil springs and direct method) for 90° elevation orientation

Description of mode	Fixed base (antenna + foundation)	With Vs 100 m/s soil springs	Foundation surrounded by soil with embedment with Vs 100 m/s
Bending in X reflector	23% 3.801 1st mode	14.28% 3.289 1st mode	~0 3.711 1st mode
Bending in y reflector	25% 4.317 2nd mode	18.5% 3.537 2nd mode	~0% 4.171 2nd mode
Torsion mode in reflector	~0% 6.699 3rd mode	~0% 6.529 4th mode	~ 0% 6.6808 3rd mode
Bending in z vertical	24.5% 18.441 11th mode	98.4% 10.55 6th mode	76% 14.699 10th mode

5 Conclusions

From the analysis carried out for soil structure interaction following conclusions are drawn.

a. In the model with embedment case with vs 100 m/s, 8.28 Hz, (X participation 82.6% - 4th mode) 8.39 Hz (Y participation 81.5% - 5th mode) and 14.7 Hz (Z participation - 76.8% 10th mode).
b. With springs, 41.8% (12.1 Hz, 7th Mode) in X, 30.3% (6.68 Hz 5th Mode) in Y, 98.4% (10.6 Hz 6th Mode) in Z.
c. For fixed base, 1st Mode is 23% (3.807 Hz bending in x) actively participated in X, 2nd mode is 25% (4.325 Hz bending in y) actively in Y 11th mode 24.5% (18.52 Hz) bending in z) in Z
d. For the case of soil springs, it is observed that, the natural frequencies are same as fixed base analyses for first 3 modes with higher shear wave velocity soil springs. For a softer founding medium, with Vs as 100 m/s, considerable reduction in natural frequencies is observed for higher modes.
e. A frequency of 10.55 Hz is obtained as a bending mode in Z (vertical) with Vs as 100 m/s, whereas, a frequency of 18.324 Hz is obtained for Vs as 760 m/s.

References

1. ASCE 4-98: American Society of Civil Engineers, Seismic Analysis of Safety -Related Nuclear Structures and Commentary
2. Indian standard IS 1893- 2002, 2016 Part 1 & Part 4 Criteria for Earthquake Resistant Design Of Structures, Part 1 - “General Provisions and Buildings”, Part 4 – “Industrial Structures including Stack like Structures”
3. FEMA 273: NEHRP Guidelines for the Seismic Rehabilitation of Buildings issued by Federal Emergency Management Agency, Washington, U.S.A., October 1997
4. Kramer, S.: Geotechnical Earthquake Engineering. Simon & Schuster, London (1996)
5. ASCE 4-16: Seismic Analysis of Safety Related Nuclear Structures, ASCE, Reston, Virginia (2016)

Author Index

C
Calçada, Rui, 54
Chang, Muhsiung, 88
Cheng, Wen-Chieh, 121

D
Dassanayake, Sandun M., 25
de Medeiros, Marcio Avelino, 13

F
Fansuri, Muhammad Hamzah, 88

G
Ghugal, Y. M., 47
Gullapalli, Venkata Lakshmi, 135

H
Hannigan, Patrick J., 64

K
Kaewunruen, Sakdirat, 54
Karande-Jadhav, Gunjan, 107
Kim, Jinkoo, 1
Kolekar, Ajay, 47
Kolhe, Pravin, 107
Konnur, B. A., 47
Kusumawardani, Rini, 88

L
Lemos, Moisés Antônio Costa, 13
Li, Ge, 121

M
Martin, Rodolfo, 54
Moghaddam, Rozbeh B., 64
Mousa, Ahmad, 25

N
Ngamkhanong, Chayut, 54
Noureldin, Mohamed, 1

P
Pradakshine, Sunil, 107

R
Rahhal, Muhsin Elie, 34
Reddy, G. R., 135

S
Satyam, Neelima, 135
Sebaaly, Graziella, 34

T
Tapase, Anand, 107
Tapase, Anand B., 47

H. El-Naggar et al. (Eds.): GeoMEast 2019, SUCI, p. 145, 2020.
https://doi.org/10.1007/978-3-030-34252-4

Printed in the United States
By Bookmasters